NAVIGATING

through

MEASUREMENT

in

PREKINDERGARTEN– GRADE 2

Linda Dacey
Mary Cavanagh
Carol R. Findell
Carole E. Greenes
Linda Jensen Sheffield
Marian Small

Carole E. Greenes
Prekindergarten–Grade 2 Editor
Peggy A. House
Navigations Series Editor

NATIONAL COUNCIL OF
TEACHERS OF MATHEMATICS

WITHDRAWN

Copyright © 2003 by
The National Council of Teachers of Mathematics, Inc.
1906 Association Drive, Reston, VA 20191-1502
(800) 235-7566
www.nctm.org

All rights reserved

Second printing 2010

ISBN 0-87353-543-x
ISBN 978-0-87353-5434

The National Council of Teachers of Mathematics is a public voice of mathematics education, supporting teachers to ensure equitable mathematics learning of the highest quality for all students through vision, leadership, professional development, and research.

For permission to photocopy or use material electronically from *Navigating through Measurement in Prekindergarten–Grade 2*, please access www.copyright.com or contact the Copyright Clearance Center, Inc. (CCC), 222 Rosewood Drive, Danvers, MA 01923, 978-750-8400. CCC is a not-for-profit organization that provides licenses and registration for a variety of users. Permission does not automatically extend to any items identified as reprinted by permission of other publishers and copyright holders. Such items must be excluded unless separate permissions are obtained. It will be the responsibility of the user to identify such materials and obtain the permissions.

The publications of the National Council of Teachers of Mathematics present a variety of viewpoints. The views expressed or implied in this publication, unless otherwise noted, should not be interpreted as official positions of the Council.

Printed in the United States of America

TABLE OF CONTENTS

CONTENTS OF CD-ROM

Introduction

Table of Standards and Expectations, Measurement, Pre-K–12

Applet Activities

How Many?
Which One?

Blackline Masters

Readings from Publications of the National Council of Teachers of Mathematics

About This Book

Navigating through Measurement in Prekindergarten–Grade 2 is the first of four grade-band books that demonstrate how some of the fundamental ideas of measurement can be introduced, developed, and extended. The introduction to this book is an overview of the development of measurement concepts from prekindergarten through grade 12. Each of the two chapters that follow the introduction focuses on a basic idea of measurement. Chapter 1 deals with comparing and ordering lengths, weights, and capacities, and chapter 2 introduces units of measure and the use of measurement tools.

Each chapter begins with a discussion of the foundational ideas and the expectations for students' accomplishment by the end of grade 2. This discussion is followed by student activities that introduce the foundational ideas and promote familiarity with them. At the beginning of each activity, the recommended grade levels are identified and a summary of the activity is presented. The goals to be achieved, the prerequisite knowledge, and the materials necessary for conducting the activities are specified. Some of the activities have blackline masters, which are signaled by an icon and identified in the materials list and can be found in the appendix. They can also be printed from the CD-ROM that accompanies the book. The CD, also signaled by an icon, contains applets for students to manipulate and resources for professional development.

All the activities have the same format. Each consists of three sections: "Engage," "Explore," and "Extend." The "Engage" section presents tasks designed to capture students' interest. "Explore" presents the core investigation that all students should be able to do. "Extend" provides additional activities for students who demonstrate continued interest and want to do some challenging mathematics. Throughout the activities, questions are posed to stimulate students to think more deeply about the mathematical ideas. After some questions, possible responses are shown in parentheses. Margin notes include teaching tips, suggestions for professional reading, and citations from *Principles and Standards for School Mathematics* (National Council of Teachers of Mathematics [NCTM] 2000). The discussion section of each activity identifies connections with the process strands and other content strands in the curriculum, offers insights about students' performance, and suggests ways to modify the activities for students who are experiencing difficulty or who are in need of enrichment. Although grade levels are recommended, most of the activities can be modified for use by students at other levels in the pre-K–grade 2 band. In order to make the modifications that will most enhance students' learning, teachers are urged to observe students' performance by taking note of the appropriateness of their mathematical vocabulary, the clarity of their explanations, the robustness of the rationales for their solutions, and the complexity of their creations.

A cautionary note: This book is not intended to be a complete curriculum for measurement in this grade band. It should, rather, be used in conjunction with other instructional materials.

Principles and Standards

Blackline Master

CD-ROM

Three different icons appear in the book, as shown in the key. One alerts readers to material quoted from *Principles and Standards for School Mathematics,* another points them to supplementary materials on the CD-ROM that accompanies the book, and a third signals the blackline masters and indicates their locations in the appendix.

NAVIGATING *through* MEASUREMENT

Introduction

Measurement is one of the most fundamental of all mathematical processes, permeating not only all branches of mathematics but many kindred disciplines and everyday activities as well. It is an area of study that must begin early and continue to develop in depth and sophistication throughout all levels of learning.

In its most basic form, measurement is the assignment of a numerical value to an attribute or characteristic of an object. Familiar elementary examples of measurements include the lengths, weights, and temperatures of physical things. Some more advanced examples might include the volumes of sounds or the intensities of earthquakes. Whatever the context, measurement is indispensable to the study of number, geometry, statistics, and other branches of mathematics. It is an essential link between mathematics and science, art, social studies, and other disciplines, and it is pervasive in daily activities, from buying bananas or new carpet to charting the heights of growing children on the pantry doorframe or logging the gas consumption of the family automobile. Throughout the pre-K–12 mathematics curriculum, students need to develop an understanding of measurement concepts that increases in depth and breadth as the students progress. Moreover, they need to become proficient in using measurement tools and applying measurement techniques and formulas in a wide variety of situations.

Components of the Measurement Standard

Principles and Standards for School Mathematics (NCTM 2000) summarizes these requirements, calling for instructional programs from prekindergarten through grade 12 that will enable all students to—

- understand measurable attributes of objects and the units, systems, and processes of measurement; and
- apply appropriate techniques, tools, and formulas to determine measurements.

Understanding measurable attributes of objects and the units, systems, and processes of measurement

Measurable attributes are quantifiable characteristics of objects. Recognizing which attributes of physical objects are measurable is the starting point for studying measurement, and very young children begin their exploration of measurable attributes by looking at, touching, and comparing physical things directly. They might pick up two books to see which is heavier or lay two jump ropes side by side to see which is longer. Parents and teachers have numerous opportunities to help children develop and reinforce this fundamental understanding by asking them to pick out the smallest ball or the longest bat or to line up the teddy bears from shortest to tallest. As children develop an understanding of measurement concepts, they should simultaneously develop the vocabulary to describe them. In the early years, children should have experience with different measurable attributes, such as weight (exploring *heavier* and *lighter*, for example), temperature (*warmer* and *cooler*), or capacity (discerning the glass with the *most* milk, for instance), but the emphasis in the early grades should be on length and linear measurements.

As children measure length by direct comparison—placing two crayons side by side to see which is longer, for example—they learn that they must align the objects at one end. Later, they learn to measure objects by using various units, such as a row of paper clips laid end to end. They might compare each of several crayons to the row and use the results to decide which crayon is longest or shortest. Another time, they might use a row of jumbo paper clips to measure the same crayons, discovering in the process that the size of the measuring unit determines how many of those units they needed. Their experiences also should lead them to discover that some units are more appropriate than others for a particular measurement task—that, for example, paper clips may be fine for measuring the lengths of crayons, but they are not practical for measuring the length of a classroom. As their experience with measuring things grows, students should be introduced to standard measuring units and tools, including rulers marked in inches or centimeters.

Children in prekindergarten through grade 2 should have similar hands-on experiences to lay a foundation for other measurement concepts. Such experiences should include using balance scales to compare the weights of objects; filling various containers with sand or water and transferring their contents to containers of different sizes and shapes to explore volume; and working with fundamental concepts of time and learning how time is measured in minutes, hours, days, and so forth—although actually learning to tell time may wait until the children are a bit older. By the end of the pre-K–2 grade band, children should understand that the fundamental process of measurement is to identify a measurable attribute of an object, select a unit, compare that unit to the object, and report the number of units. In addition, they should have had ample opportunities to apply that process through hands-on activities involving both standard and nonstandard units, especially in measuring lengths.

As children move into grades 3–5, their understanding of measurement deepens and expands to include the measurement of other attributes, such as angle size and surface area. They learn that different kinds of units are needed to measure different attributes. They realize, for example, that measuring area requires a unit that can cover a surface, whereas measuring volume requires a unit that can fill a three-dimensional space. Again, they frequently begin to develop their understanding by using convenient nonstandard units, such as index cards for covering the surface of their desks and measuring the area. These investigations teach them that an important attribute of any unit of area is the capacity to cover the surface without gaps or overlaps. Thus, they learn that rectangular index cards can work well for measuring area, but circular objects, such as CDs, are not good choices. Eventually, the children also come to appreciate the value of standard units, and they learn to recognize and use such units as a square inch and square centimeter.

Instruction during grades 3–5 places more emphasis on developing familiarity with standard units in both customary (English) and metric systems, and students should develop mental images or benchmarks that allow them to compare measurements in the two systems. Although students at this level do not need to make precise conversions between customary and metric measurements, they should form ideas about relationships between units in the two systems, such as that one centimeter is a little shorter than half an inch, that one meter is a little longer than one yard or three feet, that one liter is a little more than one quart, and that one kilogram is a little more than two pounds. They should also develop an understanding of relationships within each system of measurement (such as that twelve inches equal one foot or that one gallon is equivalent to four quarts). In addition, they should learn that units within the metric system are related by factors of ten (e.g., one centimeter equals ten millimeters, and one meter equals one hundred centimeters or one thousand millimeters). Students should clearly understand that in reporting measurements it is essential to give the unit as well as the numerical value—to report, for example, "The length of my pencil is 19 centimeters" (or 19 cm)—not simply 19.

In these upper elementary grades, students should also encounter the notion of precision in measurement and come to recognize that all measurements are approximations. They should have opportunities to compare measurements of the same object made by different students, discussing possible reasons for the variations. They should also consider how the chosen unit affects the precision of measurements. For example, they might measure the length of a sheet of paper with both a ruler calibrated in millimeters and a ruler calibrated only in centimeters and compare the results, discovering that the first ruler allows for a more precise approximation than the second.

Moreover, they should gain experience in estimating measurements when direct comparisons are not possible—estimating, for instance, the area of an irregular shape, such as their handprint or footprint, by covering it with a transparent grid of squares, counting whole squares where possible and mentally combining partial squares to arrive at an estimate of the total area. In their discussions, they should consider how precise a measurement or estimate needs to be in different contexts.

Measurement experiences in grades 3–5 also should lead students to identify certain relationships that they can generalize to basic formulas. By using square grids to measure areas of rectangles, students might begin to see that they do not need to count every square but can instead determine the length and width of the rectangle and multiply those values. Measurement experiences should also help students recognize that the same object can have multiple measurable attributes. For example, they might measure the volume, surface area, side length, and weight of a wooden cube, expressing each measurement in the appropriate units. From the recognition that multiple attributes belong to the same object come questions about how those attributes might be related. If the side length of a cube were changed, for instance, what would be the effect on the cube's volume or its surface area? Similar questions arise in comparisons between various objects. Would two rectangles with equal perimeters necessarily have the same area? What about the converse? Would two rectangles with equal areas necessarily have the same perimeter? All these measurement lessons should help students appreciate how indispensable measurement is and how closely it is tied to number and operations, geometry, and the events of daily life.

Understanding of and proficiency with measurement should flourish in the middle grades, especially in conjunction with other parts of the mathematics curriculum. As students develop familiarity with decimal numeration and scientific notation and facility in computation with decimals, applications involving metric measurements provide a natural context for learning. As students develop proportional reasoning and learn to evaluate ratios, comparisons between measurements, such as the perimeters or areas of similar plane figures, become more meaningful. Their study of geometry requires students to measure angles as well as lengths, areas, and volumes and lets students see how measurements underlie classifications of geometric figures. For example, they identify triangles as acute, right, or obtuse by evaluating measurements of their angles or classify them as equilateral, isosceles, or scalene by comparing measurements of their sides. Proportional reasoning, geometry, and measurement converge when students create or analyze scale drawings or maps. Algebraic concepts of function that develop in the middle grades have applications in relationships such as that linking distance, velocity, and time. In science classes, students use both measurement and ratios to develop concepts such as density (the ratio of mass to volume) and to identify substances by determining their densities. Through experimentation, they discover that water freezes at 0° Celsius or 32° Fahrenheit and boils at 100° Celsius or 212° Fahrenheit, and from these data they can develop benchmarks for comparing the two scales. (For example, they can see that a ten-degree change in the Celsius temperature corresponds to an eighteen-degree change in the Fahrenheit temperature or that a forecast high temperature of 30° Celsius signals a hot day ahead.)

Middle-grades students should become proficient in converting from one unit to another within a system of measurement; they should know equivalences and convert easily among inches, feet, and yards or among seconds, minutes, hours, and days, for example. They should develop benchmarks for both customary and metric measurements that can serve as aids in estimating measurements of objects. For example, they might estimate the height of a professional basketball player as about two meters by using the approximate height of a standard doorframe as a benchmark for two meters, or they might use a right angle as a basis for approximating other angle measurements like 30, 45, or 60 degrees. Although students do more computations of measurements, such as areas and volumes, during the middle grades than in the earlier years, they still need frequent hands-on measurement experiences, such as tiling a surface with square tiles, making shapes on a geoboard, or building a prism with blocks or interlocking cubes, to solidify their understanding of measurement concepts and processes.

By the time students reach high school, they should be adept at using the measurement concepts, units, and instruments introduced in earlier years, and they should be well grounded in using rates, such as miles per hour or grams per cubic centimeter, to express measurements of related attributes. As they engage in measurement activities during grades 9–12, students are increasingly likely to encounter situations in which they can effectively employ powerful new technologies, such as calculator-based labs (CBLs), graphing calculators, and computers, to gather and display measurements. Such instruments can report measurements, often with impressive precision, but students do not always understand clearly what is measured or how the technology has made the measurement. How a measurement of distance is obtained when a tape measure is stretched between two points is obvious; it is not so obvious when an electronic instrument reflects a laser beam from a surface. Thus, students need a firm foundation both in measurement concepts and in how to interpret representations of measurements and data displayed on screens.

Also during the high school years, students encounter new, nonlinear scales for measurement, such as the logarithmic Richter scale used to report the intensity of earthquakes (a reading of 3 on the Richter scale signifies an earthquake with ten times the intensity of an earthquake with a Richter-scale measurement of 2). Especially in their science classes, students learn about derived units, such as the light-year (the distance that light travels in one year, moving at the rate of $3(10^8)$ meters per second, or about 186,000 miles per second) or the newton (N) (the unit of force required to give an acceleration of 1 m/sec^2 to a mass of 1 kilogram). Students also extend ideas of measurement to applications in statistics when they measure certain characteristics of a sample and use those data to estimate corresponding parameters of a population. Students preparing for a more advanced study of mathematics begin to consider smaller and smaller iterations—infinitesimals, limits, instantaneous rates of change, and other measurement concepts leading to the study of calculus.

Applying appropriate techniques, tools, and formulas to determine measurements

To learn measurement concepts, students must have hands-on experiences with concrete materials and exposure to various techniques, such as counting, estimating, applying formulas, and using measurement

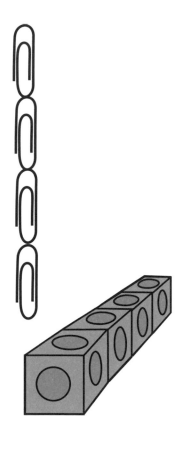

tools, including rulers, protractors, scales, clocks or stopwatches, graduated cylinders, thermometers, and electronic measuring instruments.

In the pre-K–2 years, students begin to explore measurement with a variety of nonstandard as well as standard units to help them understand the importance of having a unit for comparison. Such investigations lead them to discoveries about how different units can yield different measurements for the same attribute and why it is important to select standard units. For young children, measurement concepts, skills, and the vocabulary to describe them develop simultaneously. For example, children might learn to measure length by comparing objects to "trains" made from small cubes, discovering as they work that the cubes must be placed side by side in a straight row with no gaps, that all the cubes must be the same size (though not necessarily the same color), and that one end of the object that they want to measure must be aligned with one end of the cube train. Later, when they learn to use rulers to measure length, they must learn how to locate the zero on the ruler's scale and align it with one end of the object that they are measuring. When they attempt tasks of greater difficulty, such as measuring an attribute with a unit or instrument that is smaller than the object being measured—the width of their desks with a 12-inch ruler or a large index card, for instance—they must learn how to iterate the unit by moving the ruler or card and positioning it properly, with no gaps or overlaps from the previous position. Furthermore, they must learn to focus on the number of units and not just the numerals printed on the ruler—counting units, for example, to determine that the card shown in the illustration is three inches wide, not six inches.

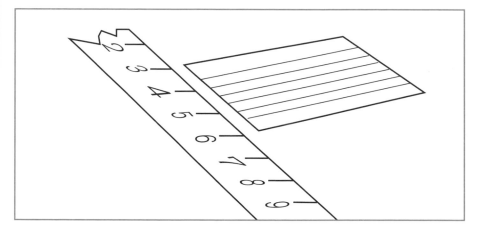

While students in prekindergarten through grade 2 are becoming acquainted with simple measuring tools and making comparisons and estimating measurements, students in grades 3–5 should be expanding their repertoires of measurement techniques and their skills in using measuring tools. In addition to becoming adept at using standard tools like rulers, protractors, scales, and clocks, third- through fifth-grade students should also encounter situations that require them to develop new techniques to accomplish measurement tasks that cannot be carried out directly with standard instruments. For example, to measure the circumference of a basketball, they might decide to wrap a string around the ball and then measure the length of the string; to measure the volume of a rock, they might submerge it in a graduated cylinder containing a known volume of water to obtain the total volume of water plus rock; to measure the weight of milk in a glass, they might weigh the empty glass as well as the glass and milk together.

As students in grades 3–5 hone their estimation skills, they should also be refining their sense of the sizes of standard units and the reasonableness of particular estimates. They might recognize 125 centimeters as a reasonable estimate for the height of a third grader but know that 125 meters or 1.25 centimeters could not be, or that a paper clip could weigh about a gram but not a kilogram. Students also should discuss estimation strategies with one another and compare the effectiveness of different approaches. In so doing, they should consider what degree of precision is required in a given situation and whether it would be better to overestimate or underestimate.

In grades 3–5, students also learn that certain measurements have special names, like *perimeter, circumference,* or *right angle;* and, as discussed earlier, they should look for patterns in measurements that will lead them to develop simple formulas, such as the formulas for the perimeter of a square, the area of a rectangle, or the volume of a cube. Through hands-on experience with objects, they should explore how different measurements might vary. For instance, by rearranging the seven tangram pieces to form a square, trapezoid, parallelogram, triangle, or nonsquare rectangle, they should find that the areas of all the shapes are the same, since they are made from the same seven pieces, but that the perimeters are different.

During middle school, students should apply their measurement skills in situations that are more complex, including problems that they can solve by decomposing or rearranging shapes. For example, they might find the area of an irregular shape on a geoboard by partitioning it into rectangles and right triangles (A) or by inscribing it in a rectangle and subtracting the areas of the surrounding shapes (B). Extending the strategy of decomposing, composing, or rearranging, students can arrive at other formulas, such as for the area of a parallelogram (C) by transforming it into a rectangle (D), or the formula for the area of a trapezoid either by decomposing it into a rectangle and two triangles (E) or by duplicating it to form a parallelogram with twice the area of the trapezoid (F). Other hands-on explorations that guide students in deriving formulas for the perimeter, area, and volume of various two- and three-dimensional shapes will ensure that these formulas are not just memorized symbols but are meaningful to them.

Students in grades 6–8 should become attentive to precision and error in measurement. They should understand that measurements are precise only to one-half of the smallest unit used in the measurement

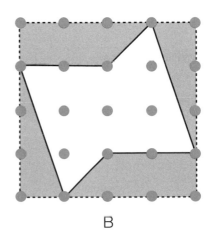

A

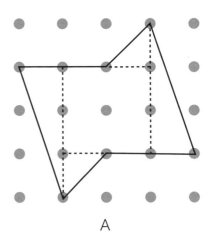

B

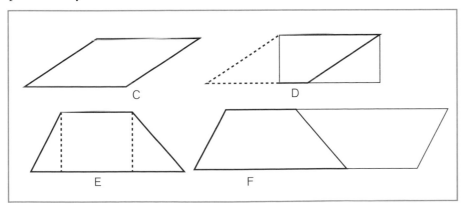

C   D

E   F

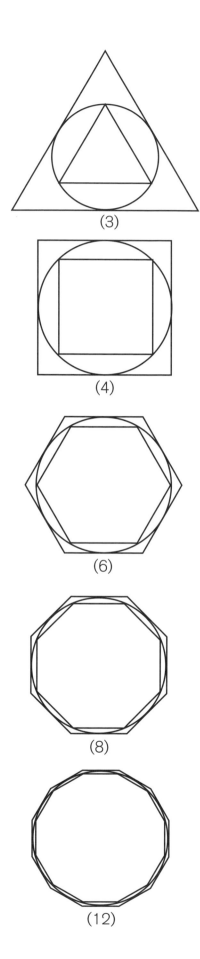

(3)

(4)

(6)

(8)

(12)

(for example, an angle measured with a protractor marked in degrees has a precision of ±0.5 degree, so a reported angle measurement of 52° indicates an angle between 51.5° and 52.5°). Students in the middle grades also spend a great deal of time studying ratio, proportion, and similarity—concepts that are closely tied to measurement. Students should conduct investigations of similar triangles to help them realize, for example, that corresponding angles have equal measures; that corresponding sides, altitudes, perimeters, and other linear attributes have a fixed ratio; and that the areas of the triangles have a ratio that is the square of the ratio of their corresponding sides. Likewise, in exploring similar three-dimensional shapes, students should measure and observe that corresponding sides have a constant ratio; that the surface areas are proportional to the square of the ratio of the sides; and that the volumes are proportional to the cube of the ratio of the sides.

Through investigation, students should discover how to manipulate certain measurements. For example, by holding the perimeter constant and constructing different rectangles, they should learn that the area of the rectangle will be greatest when the rectangle is a square. Conversely, by holding the area constant and constructing different rectangles, they should discover that the perimeter is smallest when the rectangle approaches a square. They can apply discoveries like these in constructing maps and scale drawings or models or in investigating how the shape of packaging, such as cracker or cereal boxes, affects the surface area and volume of the container. They also should compare measurements of attributes expressed as rates, such as unit pricing (e.g., dollars per pound or cents per minute), velocity (e.g., miles per hour [MPH] or revolutions per minute [rpm]), or density (e.g., grams per cubic centimeter). All these measurements require proportional reasoning, and they arise frequently in the middle-grades mathematics curriculum, in connection with such topics as the slopes of linear functions.

High school students should develop an even more sophisticated understanding of precision in measurement as well as critical judgment about the way in which measurements are reported, especially in the significant digits resulting from calculations. For example, if the side lengths of a cube were measured to the nearest millimeter and reported as 141 mm or 14.1 cm, then the actual side length lies between 14.05 cm and 14.15 cm, and the volume of the cube would correctly be said to be between 2773 cm³ and 2834 cm³, or (14.05 cm)³ and (14.15 cm)³. It would not be correct to report the volume as 2803.221 cm³—the numerical result of calculating (14.1 cm)³. Students in grades 9–12 also should develop a facility with units that will allow them to make necessary conversions among units, such as from feet to miles and hours to seconds in calculating a distance in miles (with the distance formula $d = v \cdot t$), when the velocity is reported in feet per second and the time is given in hours. Building on their earlier understanding that all measurements are approximations, high school students should also explore how some measurements can be estimated by a series of successively more accurate approximations. For example, finding the perimeter of inscribed and circumscribed *n*-gons as *n* increases (*n* = 3, 4, 5, ...) leads to approximations for the circumference of a circle.

High school students can use their mathematical knowledge and skills in developing progressively more rigorous derivations of important measurement formulas and in using those formulas in solving problems, not only in their mathematics classes but in other subjects as well. Students in grades 9–12 should apply measurement strategies and formulas to a wider range of geometric shapes, including cylinders, cones, prisms, pyramids, and spheres, and to very large measurements, such as distances in astronomy, and extremely small measurements, such as the size of an atomic nucleus or the mass of an electron. Students should also encounter highly sophisticated measurement concepts dealing with a variety of physical, technological, and cultural phenomena, including the half-life of a radioactive element, the charge on an electron, the strength of a magnetic field, and the birthrate of a population.

Measurement across the Mathematics Curriculum

A curriculum that fosters the development of the measurement concepts and skills envisioned in *Principles and Standards* needs to be coherent, developmental, focused, and well articulated. Because measurement is pervasive in the entire mathematics curriculum, as well as in other subjects, it is often taught in conjunction with other topics rather than as a topic on its own. Teaching measurement involves offering students frequent hands-on experiences with concrete objects and measuring instruments, and teachers need to ensure that students develop strong conceptual foundations before moving too quickly to formulas and unit conversions.

The *Navigating through Measurement* books reflect a vision of how selected "big ideas" of measurement and important measurement skills develop over the pre-K–12 years, but they do not attempt to articulate a complete measurement curriculum. Teachers and students who use other books in the Navigations series will encounter many of the concepts presented in the measurement books there as well, in other contexts, in connection with the Algebra, Number, Geometry, and Data Analysis and Probability Standards. Conversely, in the *Navigating through Measurement* books, as in the classroom, concepts related to this Standard are applied and reinforced across the other strands. The four *Navigating through Measurement* books are offered as guides to help educators set a course for successful implementation of the very important Measurement Standard.

NAVIGATING *through* MEASUREMENT

Chapter 1
Comparing and Ordering

"Children should begin to develop an understanding of attributes by looking at, touching, or directly comparing objects."
(NCTM 2000, p. 103)

Hiebert's (1984) discussion of children's learning of measurement concepts is available on the CD-ROM that accompanies this book.

The study of measurement should be active and involve real objects and events. Initially, young students should compare objects by lining them up to observe the differences in their lengths or heights, by holding the objects to feel the differences in their weights, and by filling two containers and transferring liquids or sand from one to the other to tell which holds more or less.

It is natural for young students to attend to the length, weight, or capacity of objects. Students' comments, such as "This is heavy," let parents and teachers know that children have begun to identify the attribute of weight and to develop a sense of the concepts heavy and light. Similarly, the comments "I'm tall," "I want the big glass," and "That took a long time" suggest that students are aware of height, capacity, and duration, respectively. Students also compare measurements every day. Such statements as "My sister is taller than you" and "My pail is bigger than yours" are examples of such comparisons.

Piaget (1960) suggested that young children should not measure length, for example, until they conserve length—that is, until they understand that the length of an object remains constant when the object is moved. Recent research, however, suggests otherwise. Hiebert (1984), for example, found that a lack of conservation did not appear to limit learning; he suggested that young children be involved in a variety of measurement experiences.

In the activities in this chapter, students compare lengths, weights, capacities, and durations. After students are able to compare two objects or events, *seriation*, the ordering of more than two objects or events by a particular attribute, is introduced. Through these introductory activities, students gain important ideas about measurement that can be applied later when they use standard units and measurement tools.

In Body Balance, students pick up objects of different weights and gain a feel for which is heavier and which is lighter. Opportunities to compare the weights of many different objects help students learn that larger objects are not always heavier than smaller ones—a common misconception among young children. Students also learn to use their bodies to simulate a pan balance. As they lean to one side to indicate a greater weight or hold both arms at the same level to indicate that two objects weigh about the same, students gain an understanding of equality and inequality. The notion of equality as a relationship, rather than as a sign that a sum, difference, product, or quotient is to be found, is an important foundation for algebraic thinking.

"Essential to developing understanding of concepts of measurement is familiarity with the language used to describe measurement relationships (i.e., longer, taller, shorter, the same length; heavier, lighter, the same weight; holds more, holds less, holds the same amount)" (Greenes 1999, p. 44). Comparative language is introduced in Scavenger Hunt. Students try to find objects that are longer than, shorter than, heavier than, or lighter than a given object or that hold more or less than an object. They then order groups of three or four objects to identify, for example, the longest and the shortest or the heaviest and the lightest. They learn that comparisons are relative: although a pencil may be longer than a crayon, the pencil may also be shorter than an umbrella.

Length is further explored in String Lengths and Ribbon Heights. Lengths may be compared in different ways. In everyday life, young students compare lengths visually. By simply looking at an automobile, for instance, they know that it is longer than a cat. When objects are similar in length, though, students must compare the lengths either directly or indirectly. In direct comparison, actual objects are placed side by side to determine which is longer, taller, or shorter or if they are the same length or height. When objects cannot be moved next to each other, they can be compared indirectly. Indirect comparison involves using a third object the same length as one of the objects being compared. In String Lengths, students estimate lengths of string that they believe will be equal to their heights, and they then compare the strings to their actual heights. In Ribbon Heights, students compare and order their heights directly, by standing back-to-back. They cut ribbons equal in length to their heights and then use the ribbons to compare their heights and to make displays of their heights from shortest to tallest. Through comparing and ordering lengths and heights, students gain insight into the need for a common end point. This initial exposure to using string or ribbon to make length and height comparisons prepares students for using a ruler later.

In Fill It Up, students compare and order containers by capacity. Although students often compare containers visually to decide which one holds more, they usually consider it necessary to compare only the heights of the containers. Direct comparisons are needed to challenge this misconception. A filler, such as sand, water, rice, or dried beans, is used to make direct comparisons. One container is filled and then its contents are poured into another to determine which one holds more and which one holds less. When students discover that a taller container holds less than a shorter one, they learn that they must pay attention to more than one dimension of the container when considering capacity.

Chronological order and duration are explored in Before, After, or Between. Students first identify and order eight events that have occurred since their births. They then informally consider the amount of time that has elapsed between two events and decide how to indicate longer and shorter durations. As they discuss events that occurred first or last or after, before, or between other events, students also develop vocabulary related to temporal events.

Temporal sequencing can be difficult for young students. They must use their memories rather than direct observation to determine what they did first, second, or third. An indication of when an event occurred may not be sufficient to stimulate their memories, so many preschool students respond, "I don't know," when they are asked what they did yesterday. Deciding which event took longer can also be difficult for students unless the events differ greatly in duration or begin (but do not end) at the same time.

Expectations for Students' Accomplishment

By the end of grade 2, students should be able to recognize the attributes of length, capacity, and weight and be able to compare and order a set of objects by each of these attributes. Students should also be able to order events chronologically and tell which event occurred first, second, third, or last; they should be able to identify events that occur before, after, or between others. Many students should be able to comment on the duration of events and tell which events take more or less time than others. Finally, at the end of grade 2, students should be familiar with and use the comparative language associated with measurement.

Body Balance

Prekindergarten–Kindergarten

Summary

Students use their bodies to indicate the relative weights of two different objects.

Recognize the [attribute] of … weight

Compare and order objects according to [weight]

pp. 68, 69

Goals

- Compare weights by feeling the difference between them
- Model a pan balance with the body

Prior Knowledge

- Identifying common objects

Materials

- A variety of objects of noticeably different weights that children can hold in their hands (e.g., a heavy book, a paperweight, a stapler, a boot, a scarf, a piece of paper, a toothbrush, paper clips); a large, lightweight object, such as a piece of plastic foam; and a small, heavy object, such as a paperweight
- A pan balance (if available)
- For each group of two or three students, three heavy items, three light items, and two items that weigh about the same
- A copy of the blackline masters "Body Balance" and "Possible or Not?" for each student

Activity

Engage

Show the students a pan balance if one is available. Identify the instrument, and tell the students that its purpose is to show which of two objects is heavier. Place a heavy book (or another heavy object) on one side and a paper clip (or another light object) on the other side of the balance. Have the students tell you how they can identify the heavier object by using the pan balance.

Hold your arms straight out and parallel with the floor, and tell the students that you are going to pretend to be a pan balance: If the objects in your hands weigh the same, your arms will stay parallel with the floor, and your hands will be even. If you hold a heavy object on one side, lean from the waist toward the side holding the heavier object, and drop one arm as the you raise the other. Dramatize a "body balance" by holding the heavy book in one hand and the paper clip in the other. Bend toward the floor on the side holding the book to show that the book is heavier than the paper clip. Talk with the students about why your body leaned over (i.e., because the book weighs more than the paper clip).

Display two objects of noticeably different weights, such as a boot and a scarf. Tell the students that you will be holding the boot in one hand and the scarf in the other. Show them which hand will hold each object. Have them predict which way your body will lean. Then pick up the two objects and lean to show which is heavier. (You should exaggerate the lean of your body to help clarify the concept.)

Ask questions like the following:

- Why did my body lean to the left (or right)?
- What would happen if I switched the scarf and the boot from one hand to the other?

Explore

Organize the students into groups of two or three, and distribute the objects. Encourage each student to pick up two objects of noticeably different weights, one in each hand, to see which feels heavier. Have the students use their bodies as balances to model the relative weights of the two objects. After the students have compared objects that differ considerably in weight, reduce the differences between the objects until the objects weigh about the same.

Initiate a discussion of what the students think makes an object heavy or light. Be prepared for misconceptions to arise. For example, some students assume that anything large is heavy and anything small is light. To help the students understand that this assumption is not always correct, let them feel a small, heavy object, such as a paperweight, and a large, lightweight object, such as a piece of plastic foam.

Give each student a copy of the blackline master "Body Balance" to complete individually. Explain that in numbers 1 through 4, the students should draw a line from each object to the hand they think is holding it. In number 5, the students should draw objects that will make the body balance correct. After all the students have finished the worksheet, talk with the class about what objects the students drew in the balance for number 5 and how they know that the objects weigh the same.

Extend

Model a "preposterous" balance situation; that is, place an object in each hand and lean toward the lighter one. Talk about why this situation makes no sense. Give each student a copy of the blackline master "Possible or Not?" After explaining the directions, allow time for the students to complete the work individually. After all the students have finished, have a class discussion about their choices.

Discussion

This activity serves as an introduction to the pan balance. Using the body as the balance is more immediately meaningful than using a pan balance because the students can actually feel the weight pulling them farther down on one side than on the other.

Initially, some students may believe that all like objects have the same weight. To correct this notion, have the students lift a variety of like objects, such as books, of different weights. For a similar demonstration, you might use plastic margarine or ice-cream tubs, leaving one empty and filling the others with different numbers of cubes or amounts of sand.

To avoid confusion about left and right, have the students face the same way that you are facing when they are predicting which way your body will lean. Alternatively, have the students simply point to the hand that will be lowered.

You may wish to duplicate the evenly balanced figure in number 5 for students who wish to draw additional situations. You might also make the items pictured actually available to the students.

Scavenger Hunt

Prekindergarten–Kindergarten

Summary

Students work in pairs to complete a scavenger hunt in which they must find objects that are longer, shorter, heavier, or lighter than other objects and objects that hold more or less than other objects.

Goals

- Identify objects that are longer, shorter, heavier, or lighter than a specified object or that hold more or less than a given object
- Understand and use comparative measurement vocabulary
- Discover the need for a common end point to compare the lengths of objects
- Order objects by a specific attribute

Prior Knowledge

- Recognizing the attributes length, weight, and capacity
- Comparing lengths, weights, and capacities

Materials

- A copy of the blackline master "Scavenger Hunt List" for each student
- A paper clip and a crayon for comparing length, a penny and a wood block for comparing weight, and a spoon and a small paper cup for comparing capacity
- Rice, dried beans, or sand for comparing volume
- Objects of various lengths, weights, and capacities for three stations where students will make comparisons

Activity

Engage

To review the comparison of lengths, weights, and capacities and the terms used to describe comparisons, ask the following questions:

- Which is longer—this paper clip or this crayon? What can you do to figure out which is longer if you aren't sure? Which object is shorter?
- Which is lighter—this penny or this block? What can you do to figure out which is lighter if you aren't sure? Which object is heavier?
- Which holds more—this spoon or this cup? What can you do to figure out which holds more if you aren't sure? Which object holds less?

Encourage the students to demonstrate how they can compare the objects.

Recognize the attributes of length, volume, weight, area, and time

Compare and order objects according to these attributes

p. 70

"Children must understand and be able to verbalize the language of comparison, for example, more *or* less, *bigger* or *smaller,* heavier *or* lighter, *and* longer *or* shorter."*
Lang (2001, p. 463)

Explore

Ask, "Does anyone know what a scavenger hunt is? Has anyone ever been on a scavenger hunt?" If the students are acquainted with this type of activity, give them time to share their stories. If they are not familiar with it, explain that on a scavenger hunt, people are given a list of things to find. You may also want to tell them that *to scavenge* means to look for and collect.

Give each student a copy of the blackline master "Scavenger Hunt List." Read the words *is lighter than a* for the first item, and ask the students to identify the pictured object. Repeat the process for each comparison, and then read the entire list several times with the students. Explain to the students that their job is to find an object that fits each of the comparisons. When they find a suitable object, they should draw a picture of it on their papers.

Organize the students into pairs, and let the hunt begin. After the students have completed the tasks, have them talk about their findings. As the students identify the objects they found, you may want them to collect the objects and demonstrate the comparisons. (For comparisons of weight, for example, the students might use their bodies to model a pan balance.)

To emphasize the importance of using a common end point when measuring lengths, hold up two objects without aligning their end points. Say, "I can't tell which object is longer. What should I do?" If no one suggests aligning the end points, explain the need to do so.

Extend

Create three sorting centers at which the students can compare various objects to a specified item. At one station, the students could determine whether objects are longer than, shorter than, or about the same length as the given item. At another station, the students could compare the weights of various objects to that of a specified object by feeling them. At the third station, the students could compare the capacities of several containers to that of a given one by filling them with sand or rice.

As the students work at these stations, observe how they make their decisions. Do they lift objects to determine which feels heavier? Do they line up objects with a common end point to compare the lengths? Do they compare amounts of rice, beans, or sand to determine which container holds more? Challenge the more able students to order the whole set of objects according to the stated attribute. Occasionally change the reference items to produce different sorting outcomes.

Discussion

Depending on the objects chosen, the students may use different methods of comparison. For example, to find something shorter than a pencil, the students might choose an object of considerably less length and simply recognize that it is shorter. If an object of similar length is chosen, the two objects may be placed side by side to allow for a direct comparison. Be sure to emphasize the importance of a common end point for comparisons of length.

String Lengths

Grades K–1

Summary

Working in groups of three, students take turns estimating their heights. One at a time, they mark the length on a string that they think will match their height. Then they compare each estimate with the actual height. The activity is then repeated using the lengths of common classroom objects instead of students' heights.

Goals

- Estimate lengths
- Compare lengths
- Discover the need for a common end point to compare lengths and heights

Prior Knowledge

- Recognizing length as an attribute of objects

Materials

- One four-foot length of string
- One six-foot length of string, thick yarn, or ribbon for each group of three students
- *Much Bigger than Martin* (Kellogg 1971)
- Tape, stickers, or a felt marker for each group of students
- Large index cards or pieces of cardboard

Activity

Engage

Have the students answer questions about relationships among the heights of their family members. Ask, "Do you have a taller brother or sister? Are you taller than someone in your family? Have you ever tried to make yourself look taller? What have you done?" (Stand on a stool, a stair, tiptoes)

Read the book *Much Bigger than Martin* (Kellogg 1971). Reinforce the idea that children are "just the right size" for their ages. Ask the students what they are able to do because they are smaller than adults (fit on a parent's lap, fit into small furniture or play areas, play on small playground apparatus, ride small bikes).

Explore

Hold up a piece of string that is about four feet long. Ask the students to stand if they think they are taller than the string is long. Use direct comparison to compare the students' heights with the length of the string. Repeat the exercise to determine which students' heights are less than the length of the string and whose heights are about the same as the length of the string.

Understand measurable attributes of objects

Understand how to measure using nonstandard and standard units

"Measurement experiences should include direct comparisons as well as the use of nonstandard and standard units. For example, teachers might ask young students to find objects in the room that are about as long as their foot." (NCTM 2000, p. 103)

To emphasize the need for a common end point, have a student stand on a stool or chair after it is determined that the student is shorter than the string. Keeping the end point of the string on the floor, again ask which is taller—the child or the string. Do not be surprised if some children answer incorrectly or are confused by this question. Some students may be distracted by the seemingly greater height of the child. Many experiences over time are needed for students to understand this aspect of measurement.

Divide the students into groups of three, and supply each group with a six-foot length of string, yarn, or ribbon. Demonstrate how you would like each student to mark the string with tape, a sticker, or a felt marker at the length that he or she estimates to be the same as his or her height. Then demonstrate how the students should compare their estimates with their actual heights.

Example

José (seated) estimates that a length of string measures the same as his height. He marks this length by placing a piece of tape on the string. He then stands to check his estimate. Another student, Alicia, holds the appropriate end of the string at José's feet, and a third student, Jamal, extends the taut string to align with the top of José's head. The students compare the length of string needed to match José's actual height with his estimate. The teacher asks, "How close was José's estimate? Is José taller or shorter than his estimate?"

Have all the students perform the estimating task with the help of the other students in the group. Encourage the estimators to verbalize the comparison in statements such as "My string length is too short. I am taller than I estimated." The students should exchange roles so that each one has the opportunity to perform each component of the task.

Extend

Call on several students to name objects in the classroom that they could measure with a piece of string. Write each item on a card (see fig. 1.1). Divide the students into pairs, and have the pairs choose one of the cards and estimate the length of the indicated item on a piece of string by marking the string with tape, a sticker, or a felt marker. If students cannot agree on their estimates, they can mark more than one

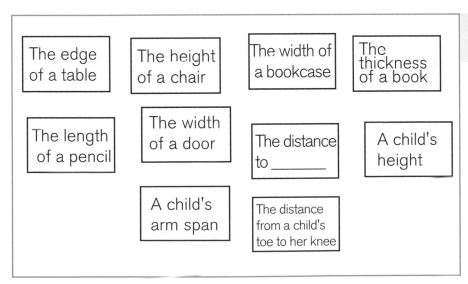

Fig. **1.1.**

Objects to measure with string

spot on the string. Have the students compare their estimates with the actual lengths of the objects. The pairs can then trade cards and measure other items.

Discussion

This estimation-and-comparison process affords students opportunities to develop their mathematical vocabulary and increases their spatial sense and powers of observation. Students learn to observe themselves and objects in their environment in order to determine similarities and differences in height or length.

Because String Lengths does not require students to make a numerical estimate, it is an excellent way of easing students into the process of estimating. As students gain greater experience with these types of estimation-and-comparison tasks, their estimates will more closely match the actual measures. Students will also learn the importance of a common starting point for comparing lengths or heights.

Fill It Up

Grades K–1

Summary

Students compare the capacities of containers and order the containers by capacity. They begin by predicting which container will hold more (or less) and check their predictions by filling the containers. They extend the comparison to three or more containers.

Goals

- Compare the capacities of containers and order the containers by capacity

Prior Knowledge

- Recognizing capacity as an attribute of objects

Materials

- Small plastic tubs (e.g., margarine or ice-cream containers) and a variety of unbreakable, translucent or transparent containers (e.g., plastic drinking glasses and empty plastic bottles) with varying capacities and shapes. Each pair of students will need four containers.

- A pitcher of water

- Materials that can be poured (water, sand, rice, or dried beans)

- A large bucket, a small plastic wading pool, a sandbox, or another large container in which materials may be poured

Activity

Engage

Gather the students around a low table on which you have placed translucent or transparent containers of different sizes and a pitcher of water. Hold up two containers with capacities that obviously differ. Ask, "Which container do you think holds more?" "Which holds less?" Once the correct containers have been identified, ask, "How could you prove that this container holds more (or less)?" Solicit several ideas from the students. If no one suggests using water, ask, "How could you use this water to help you?" Make sure that the students understand that—

- if the contents of one full container "underfill" another container, the underfilled container holds more;

- if the contents of one full container "overfill" another container, the overfilled container holds less;

- if the contents of one full container exactly fill another container, the two containers hold the same amount.

Next, show the students a tall, narrow container and a short, wide container with a greater capacity than the tall container. Ask, "Which

"Preschool children learn about volume as they pour sand or water from one container to another. In the classroom, … they also should experiment with filling larger containers with the contents of smaller ones and conjecture whether a quantity may be too much for a proposed container." (NCTM 2000, p. 104)

container do you think holds more?" If the students agree that the shorter one holds more, ask, "How can this shorter container hold more?" Elicit the explanation that the container is wider. If the students don't agree that the short, wide container holds more, use water, sand, or some other material to demonstrate that it does.

Explore

Pair the students who were successful in the activity in the "Engage" section, or pair a student who was not successful with one who was. Give each pair of students two plastic containers that have different capacities. (The capacities of the containers should not differ greatly unless you expect that the children will have difficulty with this activity.) Ask each pair of students to predict which container will hold less. The students in each pair should explain why they made their choice.

Next, give the students water, sand, rice, dried beans, or some similar material, and have them fill the container that they have chosen as the smaller one. After each pair has filled its container, ask the students to predict what will happen if they pour the contents of the container they have identified as smaller into the other one. Direct the students to pour the material from the smaller container into the larger one. Discuss whether their predictions were correct.

For this part of the activity, a classroom water table would be ideal. You might also bring a small plastic wading pool or a large bucket into the classroom to catch spills. Alternatively, you could move outside, where spilling water or another material would not be too messy.

Extend

After completing the activity with two containers, give each pair of students three containers. Ask the students to identify the containers they believe will hold the least, the most, and an amount in between. Direct the students to order the containers according to the predicted capacities. At the water table or other selected place, give them water, dried beans, or another material for pouring, and direct the students to fill the container that they have chosen as the smallest. Tell the students to pour the contents of that container into the container that they identified as having the intermediate capacity. Direct the students whose predictions were correct to fill the second container and pour its contents into the third container. Allow the students who discovered that their predicted order was incorrect to choose a new order and check the revised order by repeating the pouring process. Have the pairs repeat the ordering-and-checking process with four containers.

After all the pairs of students have successfully completed the activity, bring the entire class together, and ask the students to describe what they did to determine the order and what the results were. Ask questions like "Who has a container that will hold more than this blue glass but less than this ice-cream container?" "How might we check to see if Emma is correct?"

Discussion

For the activity in "Explore," many of your students may predict that the tall, narrow container will hold more than the short, wide one. It can be difficult for some young children to attend to two characteristics—that is, height and width. Most young children consider height dominant and thus may conclude that the taller container holds more. By presenting the counterexample, you give your students the opportunity to realize the importance of attending to both variables.

After successfully predicting and comparing the capacities of two containers and giving reasons for their choices, the students can move on to ordering three or more containers.

Ribbon Heights

Grades 1–2

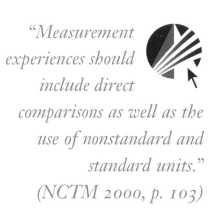

Summary

The focus of this activity is on comparing heights and lengths, both directly and indirectly. In small groups, students line themselves up from shortest to tallest by comparing their heights directly. Then, using ribbons cut to their heights, they compare and order the ribbons to determine the order of the heights of the students in their groups and in their class. They make simple vertical and horizontal displays to compare the ribbons and, indirectly, their heights. These measurement techniques are then extended to other physical characteristics.

Goals

- Compare lengths and heights directly and indirectly and order them
- Create and interpret simple horizontal and vertical displays

Prior Knowledge

- Recognizing length as an attribute of objects
- Recognizing the need for a common end point to compare lengths and heights

Materials

- A blank piece of paper for each child to be used as a recording sheet
- A long piece of ribbon or string for each child
- Scissors, pencils, and paper for each pair of students
- A large piece of butcher paper for each group of students plus an extra piece to place on the wall or on a bulletin board

Activity

Engage

Divide the class into groups of five or six students. Tell the students in each group to line themselves up from the shortest to the tallest child. Let the students decide how to determine who is the shortest, the next shortest, and so on. The students will probably look at all the students in the group to determine who is the shortest; they may use only inspection to determine how to arrange themselves. If at least some of the children are similar in height, they will need to find a way to determine which of two children is taller. They will probably have the two children stand back-to-back so that they can compare their heights directly. You may need to help the students make decisions about such matters as whether they should leave their shoes on while they are comparing heights. (The results will be more accurate if they do not.) After the students have determined the correct order, have them record it individually on their recording sheets. Note that different recordings

"Measurement experiences should include direct comparisons as well as the use of nonstandard and standard units."
(NCTM 2000, p. 103)

are possible; for example, some students might draw a picture, whereas others might make a list of the students' names.

Explore

Following the direct-comparison activity, ask a volunteer to help you demonstrate how to cut a ribbon or a string equal to his or her height. Direct the student to stand with his or her back as close as possible to a wall or a bulletin board that has been covered with a piece of butcher paper. Using a marker or a pencil, mark the student's height. Then have the student move away from the wall or the bulletin board. Have the student hold a long piece of ribbon or string against the floor directly under the mark. Unroll the ribbon, and, holding it taut, cut it where it reaches the mark.

Pair the students, and give each pair two long pieces of ribbon or string. Have the students use the method you demonstrated to cut the ribbons to the lengths of their heights. Circulate to help the students measure accurately. When they have finished, they should clearly mark their ribbons with their names.

Have the students return to their original groups and order the ribbons from shortest to longest. Allow each group to determine how best to do this task. Some groups may wish to compare all the ribbons in pairs, and other groups may realize that the order of the ribbons should be the same as the order of the heights of the students in the group. After the students have determined the order, have them tape the ribbons, ordered from the shortest to the longest, to a large sheet of butcher paper. Remind them that they should label the display by identifying the person whose height is represented by each ribbon. You may have to help them draw a line at the bottom of the paper (for a vertical display, as in fig. 1.2a) or to the right of their names (for a horizontal display, as in fig. 1.2b) to use as a baseline for the ribbons. Talk about why it is important that all the ribbons have a common baseline.

Gather the students as a class to discuss the displays they have made. Ask, "Do all the ribbon displays show the heights in the same order as the order of the lineup? Why or why not?"

If some groups have arranged the ribbons from left to right, as shown in figure 1.2a, and others have arranged them from bottom to top, as shown in figure 1.2b, call the differences to the students' attention. Ask, "Does it matter that some ribbon graphs begin at a line on the left and go to the right and others begin at a line on the bottom and go up?" Discuss with the students why these displays show the same information.

Extend

Ask the students, "How might we arrange the entire class from shortest to tallest?" They might suggest lining up the entire class or using the ribbons to find the correct order. Ask whether the order will be different if different methods are used.

Ask the students to suggest other characteristics that they might use to order themselves. Their suggestions might include the length of the right arm, the hand span, or the length of the left foot. Discuss whether the order of the students in each group will remain the same if a different characteristic is measured. Let each group decide on a new characteristic to measure, and let them determine the method of measurement to use. Have each group make a visual display of

Fig. 1.2.

Examples of ribbon displays

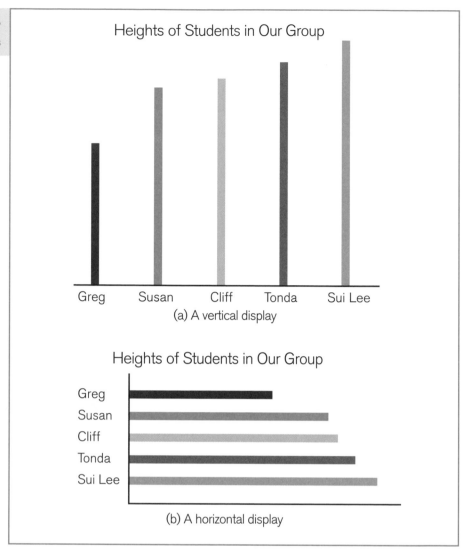

(a) A vertical display

(b) A horizontal display

the results. Ask each group to share the new information and to compare it to the information represented in the ribbon display of the groups' heights.

Discussion

This activity involves comparing students' heights directly and then indirectly by using ribbons or strings. Experiences with indirect measurement help prepare students for the use of measuring tools. When students use direct comparisons to order themselves by height, each child tends to focus on his or her own position. The use of the ribbons for indirect comparisons allows all the students to participate in the ordering of five or more lengths and thus to focus on the whole arrangement.

In this activity, the students also develop skills in displaying the data that they collect. Furthermore, they compare different representations of data and look for relationships between two sets of data.

Before, After, or Between

Grades 1–2

Summary

Students identify eight events that have occurred since their births, draw pictures of the events, and label the pictures. Each student attaches his or her pictures to a clothesline in the order in which the events occurred. The discussion focuses on temporal comparisons (i.e., what happened first, last, before, and after) and on the spacing of the pictures to represent the actual intervals of time accurately.

Goals

- Order events according to the time they occurred
- Identify events that occur *before*, *after*, or *between* other events
- Construct time lines to show the order of events and the time between events

Prior Knowledge

- Using the ordinal vocabulary *first, second, third, …,* and *last* to describe the order in which activities occur in a day

Materials

- Three yards of clothesline or heavy string or yarn for each student
- Eight clothespins or large paper clips per student plus ten extra ones
- Drawing paper and markers or other drawing implements

Activity

Engage

Call on students to tell what they did during the day from the time they awoke in the morning to this discussion. Emphasize the use of ordinal language (e.g., *first, second, third*) and sequential language (e.g., *next, before, after*). The students might want to record their day's events and talk about how their lists are similar and different.

Explore

As a homework assignment, have the students ask their families to help them identify eight important events in their lives, beginning with their births. Have the students bring in their lists of events with the years and, if possible, the months identified.

Have the students draw pictures of each event on a separate piece of paper and label each picture with a brief description and the date of the event. After all the pictures have been completed, give each student a clothesline and clothespins, and have the students attach the pictures of the events to the lines in the order in which the events occurred.

Lay the time lines on the floor. Call on students to describe the events in their lives, beginning with the earliest one. You may want to have the students swap time lines and tell each other's "life story."

Recognize the [attribute] of … time

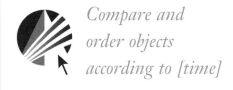

Compare and order objects according to [time]

Although your students have seen the four-digit numbers for years, they may not have learned that the smaller of two numbers refers to an earlier date. When the students compare their ages, they may not initially understand that the smaller number for the year of birth represents the older student. A discussion about this topic is important, and the students will learn from making many comparisons of dates.

Ask questions like the following to promote an understanding of the order of events and the relative positions of the events on the time lines:

- What happened on __(date)__ ?
- What was the date of the __(event)__ ?
- Which happened *first*— __(event)__ or __(event)__ ? How do you know?
- What happened *between* __(event)__ and __(event)__ ?
- What happened *before* (or *after*) __(date)__ ?
- What happened *before* (or *after*) __(event)__ ?

Be sure that the students understand that the words *before, after,* and *between* are normally used when more than one event is involved and that *between* does not include the events at the end points. Display the "life lines" by hanging them on a wall or by suspending them from the ceiling.

Extend

Construct a time line describing ten events in *your* lifetime. Use one sheet of paper for each event and a twenty-foot piece of clothesline. Show only the years in which the events occurred, and select events that are not separated by equal intervals of time. Have the students help you clip the events to the clothesline in the proper order.

Call on a student to tell your "life story" and identify the year that each event occurred. Discuss various ways to indicate more or less time between two successive events than between two others. Using the students' ideas, rearrange the pictures to reflect the time between all successive events more accurately.

If the students' time lines do not accurately represent the time between events, suggest that they revise their time lines. The activity can be extended to constructing time lines of important events in the lives of older relatives or friends or of historical figures.

Discussion

Before, After, or Between not only offers opportunities for students to explore the order of events over time but also introduces the concept of elapsed time. That concept and ways of determining elapsed time are developed in grades 3–5. Before, After, or Between also reinforces students' understanding of the calendar.

As they tell their life stories or the stories of others, students develop their oral storytelling abilities. As they write descriptions of events in their lives, read descriptions written by their peers, and interview adults, they further develop their literacy skills. Finally, illustrating the events enhances students' artistic abilities.

Conclusion

This chapter has introduced students to the concept of measurement through activities involving comparing and ordering actual objects by length, weight, and capacity. Chapter 2 introduces students to standard and nonstandard units of measure and to the use of tools for measuring.

The time line is like a number line with the distances between the pictures representing the time between events. In the introductory activity in "Explore," the students will likely not attend to such differences but focus only on ordering the pictures. In the extension activity, they should attend to the relative elapsed times.

NAVIGATIONS
SERIES

PRE-K–GRADE 2

NAVIGATING *through* MEASUREMENT

Chapter 2
Using Units and Tools

"Length should be the focus in this grade band, but weight, time, area, and volume should also be explored."
(NCTM 2000, p. 44)

Once students have developed a well-grounded understanding of length, weight, and capacity through comparison activities, they are ready to explore the use of units of measure and measuring tools. Through explorations with a variety of nonstandard and standard units, students begin to generalize the measurement process as assigning a numerical value to a particular attribute of an object or to an event.

Linear measure is the most visible attribute of an object. Initially, young students determine length by using multiple copies of a unit. The units are placed end to end until they line up to approximate the length of the object. Young children do not, however, necessarily attend to the positions of the units. The units may be placed so that they overlap, so that there are gaps between them, or so that they form a crooked arrangement. Although such placements will yield imprecise measurements, such behaviors are normal for young children. Over time and with greater discussion about, and exposure to, measurement, students will learn to be more precise.

In Giant Steps, Baby Steps, students measure a designated distance in steps of different sizes. Young children are familiar with taking baby steps and are likely to place a heel directly in front of a toe. Thus overlaps and gaps are minimized. Significantly different numerical values will be assigned to an object when the different units (giant steps and baby steps) are used to measure, so the activity offers children the opportunity to think about the inverse relationship between the value assigned and the size of the measurement unit. Specifically, they find that when longer units are used, the number of units is less and that when shorter units are used, the number of units is greater.

"Recent research suggests that constructing measurement tools can contribute to students' understanding of concepts of measurement."
(Wertsch 1998)

"If students initially explore measurement with a variety of units, nonstandard as well as standard, they will develop an understanding of the nature of units."
(NCTM 2000, p. 105)

In Snake Imprints, students use multiple copies of a unit and then use one unit repeatedly (iteration) to measure the lengths of objects. Using one unit repetitively in the iterative process can be problematic. Students forget or ignore the location of the end of the previous placement. In this activity, a snake, equal in length to the object being measured, is constructed of modeling dough or clay, and the students make impressions of the single unit along the snake. This technique allows young children to "see" the unit after it has been removed and facilitates the counting process. As an extension to this activity, students make simple rulers by joining twelve one-inch strips.

Snail Trails focuses on ways to find the measures of nonlinear objects. Students use string to copy a crooked path and the distance around an object, and they then measure the string with nonstandard or standard units. Students are introduced informally to the idea of perimeter as they find the distance around rectangular figures by combining the lengths of the sides. Note that the use of standard and nonstandard units is combined in both Snake Imprints and Snail Trails.

Although in many activities in this chapter students are encouraged to estimate before actually measuring, the primary focus of Estimation Challenge is the estimation of length in standard units. Estimation, which is further developed in How Many in a ___?, helps students become more familiar with standard units and is a highly practical skill. It involves both spatial reasoning and number sense.

As with computation, students develop a variety of estimation techniques, but they require rich experiences over time to do so. Sometimes a known measure (referent) is used to estimate a measure that is not known. For example, knowing that a particular doorway to a room is seven feet tall makes it easier to estimate the height of the wall next to that doorway. Lengths can also be partitioned to aid in estimation. Knowing that about half the length of a particular bookcase is six pencils makes it easier to estimate the entire length of the bookcase.

It is not uncommon to see children and adults unitize the length of an object by estimating the length of an inch or a foot with their hands and then determining how many of those estimated units are equal in length to the object. The use of this technique, however, requires a sense of the length of an inch or a foot. The use of benchmarks for standard units, such as using the width of a child's finger to approximate one centimeter, helps students develop their sense of the units.

The activities Which Unit Did I Use? and Grandma both focus on the importance of the size of standard units of length. In Which Unit Did I Use? students are given numbers representing the lengths of a variety of objects without the units indicated. By examining the objects, the students must determine which metric unit (centimeter, decimeter, or meter) was used. They also make decisions about which units to use to measure different lengths. The activity Grandma emphasizes the importance of using standard units to communicate information about length. This idea also reinforces the notion that the same unit must be used in comparing measurements. The humor embedded in the activity will help children grasp this important concept.

Scoop It! focuses on volume, and Balance the Pans focuses on weight. Both are designed for young children but can easily be adapted for primary-grades students. In both activities, students use nonstandard units to measure. Rice or dried beans are used to measure capacity, and cubes or washers are used to measure weight. In both cases, students use measurement data to compare and order objects. The use of units of different sizes to measure weight and capacity and the relationship of the number of units to the size of the unit are explored as well. These explorations with informal units prepare students for the later use of standard units of weight and capacity.

Area is explored in Cover Up. Area is an important measurement topic, even for young children. Initially, students compare areas by superimposing one figure on another or by cutting the pieces of one figure to fit on top of another. This comparison process associates area with "covering space" and helps students differentiate area from length. Although some students may choose to iterate a single unit, this activity involves the use of multiple units. Research shows that some children move a single unit in a regular and systematic way while attempting to count the number of times it fits into a larger shape. Other children move it around randomly and unsystematically. Still others move the unit around the perimeter of the larger shape but ignore the inner part of the shape (Outhred and Mitchelmore 2000).

Calendar Logic and Fit the Facts involve mathematical reasoning about measurement. In Calendar Logic, students are given information about the days of the week and the dates in a month, as well as temporal and ordinal terms, and must reason deductively to identify a particular day and date. This activity helps students think about the relationships between days of the week and dates and develops their mathematical vocabulary.

Fit the Facts presents students with incomplete measurement stories. Students must choose number "facts" from a given list and place them in the story so that the measurements make sense. This activity stimulates mathematical reasoning and communication and helps students focus on the sizes of standard units.

Expectations for Students' Accomplishment

By the end of grade 2, students should recognize that length, area, capacity, weight, and time can be measured. They should recognize that both a number and a unit are needed to report a measure and that the size of the unit is related to the size of the number.

Although the measurement activities for the younger students in this grade band involve using several copies of the same unit (as opposed to iterating one copy), by the end of grade 2, students should also be able to use one unit repeatedly and to use a ruler to measure lengths. Similarly, although many measurement activities used in this grade band involve nonstandard units, by the end of grade 2, students should also be able to use standard (metric and customary) units to measure lengths.

Giant Steps, Baby Steps

Prekindergarten–Kindergarten

Summary

Students measure the same distances in both baby steps and giant steps. As they count and compare the number of baby steps and giant steps, they learn that the smaller units require a greater number of steps.

Goals

- Use nonstandard measurement units
- Discover the relationship between the size of a measurement unit and the number of the units required to measure the length of an object

Prior Knowledge

- Counting to 30

Materials

- Masking tape to mark start and finish lines
- Chart paper and a marker or a chalkboard and chalk

Compare and order objects according to [length]

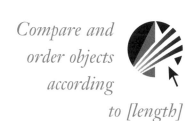

Activity

Engage

Use masking tape to mark a start line and a finish line on the floor of the classroom. The distance between the two lines should be about ten feet. Call on a student to demonstrate a giant step and then take giant steps from the start line to the finish line. Demonstrate how to take a baby step by placing your feet heel to toe. Call on a student to take baby steps from the start line to the finish line.

Explore

On chart paper or the chalkboard, make a chart listing the names of the students in your class and showing three columns, one each for number of baby steps, of giant steps, and of normal walking steps (see fig. 2.1). The latter will be used in the "Extend" part of the activity. Have the students work in pairs. While one student "steps off" the distance, first using giant steps and then using baby steps, the other student counts the steps and records the number of steps on the chart. If the students are not able to record the results on the chart, you should do it for them.

When all the students have finished, bring the class together to compare the numbers of giant steps and baby steps. Facilitate a discussion about why they took fewer giant steps than baby steps or more baby steps than giant steps to cover the same distance.

Extend

Point to the third column on the chart. Call on a student to demonstrate a walking step. Call on other students to explain how walking

Stepping Chart			
Students	Measure in Baby Steps	Measure in Giant Steps	Measure in Walking Steps
Ang			
Anya			
Brooke			
Cal			
Casey			

Fig. **2.1.**

A chart on which to record measures of a distance in baby steps, giant steps, and walking steps

steps differ from baby steps and giant steps. Ask the students to imagine that they are taking normal walking steps to walk from the start line to the finish line. Ask, "Do you think you will take more walking steps than giant steps?" "Why?" "Do you think you will take more walking steps than baby steps?" "Why?"

Have the children take walking steps from the start line to the finish line. Record the number of steps in the chart, and talk about whether the speculations matched the results.

Discussion

In Giant Steps, Baby Steps, students are introduced to the inverse relationship between the size of a measurement unit and the number of the units required to measure a length. This relationship is complex, but with frequent exposure, even very young students can come to understand it. Numerous instances of this inverse relationship occur in mathematics, particularly in dealing with measurement and money. For example, suppose that you have $5 and want to purchase items that cost the same amount. The greater the price of the items, the fewer of them you can buy for $5. Likewise, the less the price of the items, the more of them you can purchase.

By imagining the relationship between the number of walking steps and the number of baby steps or giant steps, students make estimates and predictions on the basis of established data. It is important that the students provide rationales for their speculations.

Balance the Pans

Grades K–1

Summary

Students use units such as cubes and washers to determine the weights of a variety of lightweight objects. They then compare and order objects by weight and identify sets of objects that weigh about the same.

Goals

- Use nonstandard units and a pan balance to determine the weights of objects
- Compare the weights of objects and order objects by weight
- Use appropriate vocabulary to describe the weight relationships among objects

Prior Knowledge

- Recognizing weight as an attribute of objects
- Using a pan balance to compare the weights of two objects
- Counting to 20
- Writing numerals to 20

Materials

- Six collections of a variety of lightweight objects (a box of paper clips, a pencil, a crayon, a spoon, a cup) and one heavy object (a jar of paint, a book, a box of cubes)
- Six pan balances
- Six collections of washers or cubes to use as measurement units
- Tubs of tiles, counters, cubes, washers (or other uniform lightweight objects)
- *Just a Little Bit* (Tompert 1993) (optional)
- Paper and pencils

Activity

Engage

If *Just a Little Bit* (Tompert 1993) is available, read it to the students. The story begins with an elephant and a mouse sitting at opposite ends of a seesaw, with the mouse's end staying up in the air. One by one, several animals come to join the mouse, but the seesaw still doesn't move. Finally, the mouse and its companions get the extra little bit of weight they need when a beetle lands on the mouse's nose. The seesaw moves down, and the animals can then seesaw properly.

After reading the story, encourage the students to talk about how they think a seesaw works. Invite a student to model (with his or her arms) a seesaw with a mouse on one side and an elephant on the other

Compare and order objects according to [weight]

Use tools to measure

side. Have other students identify the side with the mouse. Ask, "Why did the seesaw move when the beetle landed on the mouse's nose?"

Display a pan balance, and ask if any of the students remember what it is called. Show them how to adjust the scale when empty pans fail to balance, and explain that scales may become out of balance with use. After you have adjusted the scale, point out that the empty pans are in balance, and ask the students to tell you why it is important to start with the pans balanced.

Hold up two items of different weights, such as a jar of paint and a pencil. Place the jar above one pan and the pencil above the other pan. Have the students predict how the pans will look after you lower the objects into them. Place the items in the pans to check the students' predictions, and ask, "What does the position of the pans tell us about the weights of these objects?" If you read the story, ask, "Does this scale remind you of anything in the mouse-and-elephant story?"

Explore

Place one of the objects—for example, a box of paper clips—in one pan, and ask, "How many of these washers (or other nonstandard units) do you think we should put in the other pan to make the scale balance?" After the students have made their predictions, balance the scale by placing washers in the pan, one at a time, as the students count them aloud. It is unlikely that the pans will balance exactly; the aim, rather, is to conclude that the weight of the box of paper clips is "about the same" as a certain number of washers.

Remove the paper clips, and place a spoon in the pan. Again have the students predict the number of washers that will be needed to balance the scale and then count aloud as you place the washers in the pan. Ask, "Is the box of paper clips or the spoon heavier? Why do you think so?"

Divide the students into six groups. Have the students, working in their groups, use washers (or other units) to determine the weight of the various objects. As they work, the students should record their findings. Finally, challenge the students to use these data to order the objects by weight.

Extend

Give the students additional opportunities to use the pan balance by creating a "weight" activity center supplied with tubs of uniform light-weight objects. The students can continue to weigh the various items but with different units, such as tiles or counters. Encourage the students to discuss why more of some units than of others are needed to balance the scale.

Challenge the students to balance sets of different units. For example, the students can determine how many fresh pencils it takes to balance six washers and then to balance six cubes.

Discussion

As the students work in small groups, you will have an informal opportunity to observe their abilities to measure weight. As you move from group to group, note the following:

- At the start, do the students check to make sure that the empty pans are balanced?

"Young children should also have experiences with weighing objects. Balances help them understand comparative weights and reinforce the concept of equality; for example, they can predict that two cubes will weigh the same as twenty links and then test their prediction."
(NCTM 2000, p. 104)

- Do the students recognize that they should add weights to the higher pan in order to balance the scale?
- Do the students recognize when the pan has been balanced?
- Do the students recognize that the heavier the measurement unit, the fewer the units needed to balance the pans?

After the students have completed several tasks with the pan balance, you may want to observe them again to gain a sense of the growth of their understanding of weight.

Snake Imprints

Grades K–1

Summary

Students make the transition from using many copies of a nonstandard unit to measure length to using one copy of a unit repeatedly. Students also make simple rulers.

Goals

- Measure length using nonstandard and standard units
- Use iteration of a single unit rather than multiple copies of the unit to measure length
- Make a simple ruler

Prior Knowledge

- Recognizing length as an attribute
- Using multiple copies of a unit to measure length

Materials

- Modeling-dough snakes for each pair of students
- A set of fifteen or twenty large paper clips or pencils
- One of each of the following items for each pair of students: a toothpick, a short pencil, a paper clip, a Cuisenaire rod, and a plastic link
- Twelve narrow, one-inch strips of paper—six each of two different colors for each student
- One twelve-inch strip of tagboard per student
- Paste or glue

Activity

Engage

Gather the students in a circle near your desk, a table, or another long object in the classroom. Invite one student to measure the length of the object by using multiple copies of a nonstandard unit, such as a large paper clip or pencils of equal length. Have the students count the number of units placed end to end and report the length of the object. Discuss with them why it is important to ensure that each new paper clip or pencil be placed exactly at the end of the previous one, without leaving gaps or overlaps. To reinforce that point, you might first place the pencils or paper clips haphazardly across the object.

Use modeling dough to make a "snake" the length of the object (see fig. 2.2). Place it on the edge of the object, and starting at one end, use the same pencil or paper clip to make an impression in the snake. Move the unit repeatedly, each time making a new impression at the end of the previous one. Once the entire object has been measured by the impressions, have the students again report its length.

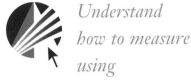

Understand how to measure using nonstandard … units

Measure with multiple copies of units of the same size

Use repetition of a single unit to measure something larger than the unit

Fig. **2.2.**

Measuring with iterations of nonstandard units

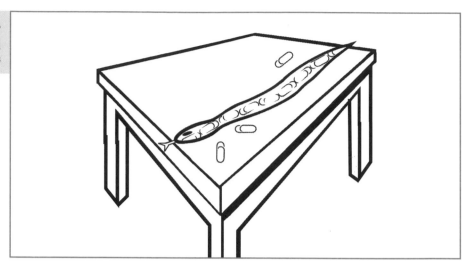

Have the students discuss the following:

* Why the number of impressions is identical to the number of units used earlier
* Why it might be easier to use a single copy of a unit repeatedly than many copies of the unit
* How using the modeling dough helped them use a single copy of a unit

The students who are able to do so could write about their use of the modeling dough. (See the students' explanations of the advantages of using modeling dough in figs. 2.3 and 2.4.)

Fig. **2.3.**

A student's explanation of the advantages of using the modeling-dough snakes

Fig. **2.4.**

Another student's explanation of the advantages of using the modeling-dough snakes

Explore

Divide the students into pairs, and give each pair (1) modeling dough from which to make snakes and (2) one each of several nonstandard units, including toothpicks, short pencils, Cuisenaire rods, plastic links, and paper clips. Assign four classroom objects of varying lengths to be measured. Have the students make snakes, lay them on the object to be measured, and make impressions of the nonstandard unit in the dough snakes. Direct them to reroll their snakes after each use to clear the impressions left on the snakes by the previous measuring unit. Encourage the students to estimate the length before they measure. Observe the students to ensure that they place the units properly along the snakes. Have each pair of students complete a chart to report their findings.

Extend

In this extension, the students focus on the use of standard units to measure the lengths of objects. Have the children, using the one-inch strips you have prepared, measure the lengths of their pencils and other objects that are less than twelve inches long. Then distribute the twelve-inch strips of tagboard, and have the students paste their one-inch strips, in alternating colors, along the edge of the tagboard strip (see fig. 2.5). Fastening the individual units together helps the students connect the actual units with the spaces on a ruler. Alternating the colors of the strips helps students identify and count the separate units in the continuous configuration. No numbers should be marked in this initial, simple ruler because numbers tend to draw students' attention to the end points of the units rather than to the intervals.

Have the students use this tool to remeasure the four objects they measured earlier. Discuss with the class the advantages of using units that have been attached end to end.

It is not necessary for each student to measure each object with all the nonstandard units. The goal is for students to realize that any of the units can be iterated. Just make sure that the students use at least two of the units in their measuring activities.

 The applet How Many? on the accompanying CD-ROM offers students practice in estimating in nonstandard units and then measuring to check their estimates.

Measures of Objects

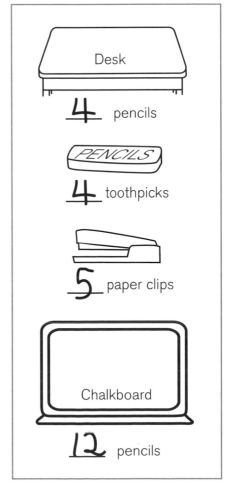

A Sample Chart

Fig. **2.5.**

A student-made ruler

Discussion

The transition from using many copies of a unit to iterating the unit to using the snake on which the unit has been iterated is a natural progression. Before students can make sense of using a ruler, it is essential that they become proficient at using a single unit (e.g., an inch or a centimeter) repeatedly to measure the lengths of objects. Making a ruler helps students see the embedded units as lengths rather than focus on the end points indicated by numbers. For students who do not have modeling dough or clay available at home, you may wish to set up a modeling center so that they can become familiar with it before they participate in the activity.

During the activity, some students may discover that the longer the unit of measure, the fewer the units necessary to "measure" the length of an object. Take advantage of that discovery to ask the students to predict the number of units of different sizes that will be required to measure the length of an object.

Scoop It!

Grades K–1

Summary

Students measure the capacities of containers by counting the number of scoops of rice or beans required to fill each container. They compare the capacities of containers by comparing the numbers of scoops needed to fill the containers. They then order the containers by capacity and explore the relationship between the size of a scoop and the number of scoops.

Goals

- Use nonstandard units (scoops) to quantify and compare capacities
- Compare the capacities of containers to tell which holds *more* and which holds *less* or if they hold the *same* amount
- Order containers by capacity
- Identify the relationship between the sizes of scoops and the number of scoops needed to fill a container

Prior Knowledge

- Counting to 20
- Recording numerals to 20

Materials

- A large open container of dried rice or beans for every four to six students
- Scoops of the same size (for laundry detergent, ice cream, coffee, etc.)—two for each pair of students
- Six scoops of other sizes
- Containers of varying sizes (unbreakable jars, bottles, vases, pails, measuring cups, etc.)—two for each pair of students. Use permanent ink to label each container with a different letter of the alphabet.
- Pencils and paper
- A chalkboard or other classroom board

Activity

Engage

Show the class two containers, one taller than the other but with less capacity (see fig. 2.6). Label one container "A" and the other "B." Call on a student to identify the letters on the two containers. Referring to the containers by letter, ask the students to tell you which one holds more and how they know that it does. Once the students have expressed several opinions, talk about how the capacities can be compared. If the students suggest ways to fill and compare the containers, follow their suggestions. Be sure that by the end of the discussion, you have

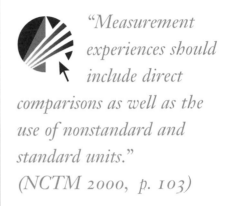

"Measurement experiences should include direct comparisons as well as the use of nonstandard and standard units."
(NCTM 2000, p. 103)

demonstrated that comparisons can be made by filling one container and pouring the contents into the other container. If the second container has not been filled, then the second container holds *more*. If the second container overflows, then it holds *less*. Follow the same procedure with other pairs of containers. Begin with pairs of containers whose capacities are decidedly different, and progress to pairs of containers with equal or nearly equal capacities.

Fig. **2.6.**

Two containers with different capacities

Explore

Demonstrate how a scoop can be used to measure the capacity of a container. Using one of the identical scoops, fill one container with rice as the students count the number of scoops. Record the capacity of the container on the chalkboard (e.g., "A holds 4 scoops").

Divide the students into pairs, and give each pair two of the identical scoops and two containers. Have the students measure the containers and record the measurements. After the pairs have completed their work, have them swap containers with other pairs of students and measure the new containers.

Most containers do not hold whole numbers of scoops of rice or beans. The students will recognize this problem. Encourage them to offer suggestions about how to record fractions of scoops.

After all the pairs have measured six containers, have them compare their measurements. If any major discrepancies are reported, have the students discuss how they used the scoops. For example, did they round the scoops or level them? You could have the students agree on a technique and then remeasure the containers. Use the data collected to compare the capacities of the pairs of containers.

Show the class groups of three or four containers. Tell them that their job is to line up the containers in order, with the one that holds the least at one end and the one that holds the most at the other end. Allow the students to decide whether to measure each container by filling it with scoops of rice or to use the measurements already recorded.

Extend

Show the entire class two scoops of different sizes. Call on a student to use the smaller scoop to fill one of the containers and determine its capacity in small scoops. Call on a different student to use the larger scoop to fill the same container and determine its capacity in large scoops. Record both measurements on the chalkboard. Ask the students why the numbers of scoops differ. If the students do not comment on the sizes of the scoops, ask them why it took more small scoops than large scoops to fill the container. Alternatively, the students who are able to do so can write about their ideas. (See one student's explanation in fig. 2.7.)

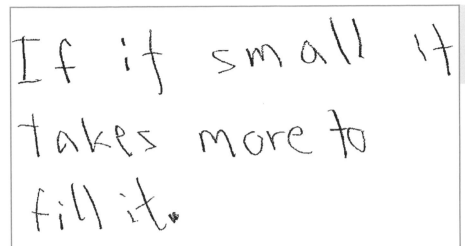

Fig. **2.7.**

A student's explanation of the relationship between the size of a unit and the number of units necessary to fill a container

If it small it takes more to fill it.

Follow the same procedure using two other pairs of scoops. Ask the students to predict—before the containers have been measured—whether more large scoops or more small scoops will be needed to fill the containers. They can then check their predictions by measuring. Some students may recognize the inverse relationship between the size of the scoop and the number of scoops required to fill a container—that is, the larger the scoop, the fewer scoops required, and the smaller the scoop, the more scoops required. Follow up by discussing why it is important to use scoops of the same size when comparing the capacities of containers.

Discussion

In any measurement activities, the power of using uniform measurement tools should be stressed. Students should realize that using uniform scoops of the same filler makes it easier to compare capacities and order containers by capacity. The example of scoops of different sizes producing different numerical measures for the same container establishes the need for standard units of measure. This concept is further developed in grades 3–5.

When your students understand the ideas developed in this lesson, you could introduce the eight-ounce measuring cup as a unit used to measure larger containers. Point out that the ounces are marked on the cups, and explain that the marks are useful for measuring quantities that are less than a cup.

Since some fillers pack more tightly than others, the same filler should be used when comparing the capacities of containers.

Snail Trails

Grades 1–2

Summary

Students explore ways to estimate and find the lengths of "crooked paths," and they use string to find the distances around objects. As a challenge, students find the perimeters of rectangular shapes by combining the lengths of the sides.

Goals

- Estimate and measure the lengths of irregular paths
- Estimate and measure the distances around objects
- Find the distances around rectangular shapes by combining the lengths of the sides

Prior Knowledge

- Measuring lengths with nonstandard and standard units

Materials

- One six-foot length of string or thick yarn for each pair of students
- A ruler for each pair of students
- Square tiles
- Cubes for measuring lengths and representing snails
- Live snails for observation (optional)
- *The Snail's Spell* (Ryder 1992) (optional)

Activity

Engage

Read the following description aloud. Ask the students to model the description (see fig. 2.8) and think about the animal that you are describing.

> You are crouched down on the ground.
> You are getting smaller, smaller, and smaller.
> You are two inches long.
> You have two antennae.
> You carry your house on your back.
> You move by gliding. You ride on your own smooth path.
> Your feelers touch everything around you as you glide.

Ask the students to identify the animal (a snail). If live snails are available, have the students observe them and compare them with the description.

Explore

Have the students, working in pairs, take turns pretending to be a snail that leaves a crooked trail and using a piece of string or yarn to

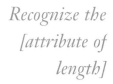

Recognize the [attribute of length]

Understand how to measure using nonstandard and standard units

Measure with multiple copies of units of the same size, such as paper clips laid end to end

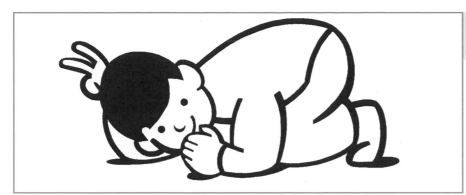

Fig. **2.8.**

A student imitating a snail

trace the "snail's" trail (see fig. 2.9). After about a minute, have the students look at the trail and estimate its length in feet. Using a ruler, have them measure the trail to the nearest foot. Note whether they measure the trail as it is laid out or the string as it is held taut. Ask questions that guide the students to understand that the original path is the same length as the straightened trail.

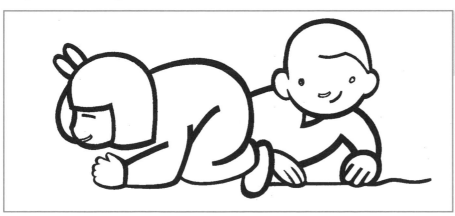

Fig. **2.9.**

One student imitates a snail while a second student marks the "snail's" trail with string

Repeat the task with objects such as cubes representing the snails. Have the students trace the small snail trail with string, estimate its length in cubes or inches, and then use cubes or a ruler to measure the trail. (See fig. 2.10.)

Fig. **2.10.**

A student measuring a "snail trail" with cubes

The number of units can be determined by iteration or by using a measurement tool.

In *The Snail's Spell* by Joanne Ryder (1988), a sleeping boy shrinks to snail size and experiences a snail's garden world.

Fig. **2.11.**

A rectangular arrangement of tiles with a perimeter of ten units

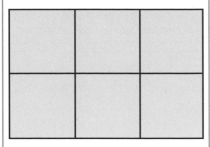

Invite the students to suggest a garden fruit or vegetable that a snail might crawl around (tomato, cucumber, squash, cantaloupe, watermelon). Have the students, working in pairs, find a round object in the classroom and pretend that the object is a fruit or a vegetable that the snail has crawled around. The students' task is first to estimate and then to measure how far the snail traveled around the object. They should cut a length of string equal to the distance around the object and then measure the string to the nearest inch or foot.

If *The Snail's Spell* (Ryder 1988) is available, read it to the students.

Extend

Have the class explore ways to find the distance around a rectangular object. Give the pairs of students ten, twelve, fourteen, sixteen, or eighteen square tiles. Have them use the tiles to make rectangles and then estimate before finding the distance around the rectangles, counting each exposed side of a tile as a unit. Ask the students if they can figure out the distance around the rectangle without counting each unit. Guide them to add the number of units on the sides. For example, the rectangle shown in figure 2.11 would be 2 + 3 + 2 + 3, or 10, units around. Challenge the students to devise ways to find the distance around rectangular objects (e.g., book or magazine covers or posters or calendar pages) in the classroom.

Discussion

When the students are using rulers, be sure that they understand that one end of the string they are measuring should be aligned with the zero mark on their rulers. Since many rulers do not have the zero recorded, point out that the first mark at the left-hand end of the ruler is the zero mark. In the upper grades, students can learn to use a ruler starting at any point.

It is important that students have numerous experiences finding distances around objects of different shapes. This practice sets the stage for the later development of formulas for the perimeters of polygons and the circumferences of circles.

Estimation Challenge

Grades 1–2

Summary

In Estimation Challenge contests, groups of students compete in estimating lengths in inches or feet.

Goals

- Estimate the lengths of objects to the nearest inch or foot

Prior Knowledge

- Measuring from one to twelve inches to the nearest inch

Materials

- A copy of the blackline master "Estimation Challenge" for each student
- Gold, silver, and bronze medals prepared from the drawings of medals on the blackline master "Estimation Challenge Medals." The medals are awarded to every student in a group. Be prepared with extra gold and silver medals in case of a two- or three-way tie.
- Twenty-four-inch ribbons for hanging the medals
- Objects of various lengths that are less than twelve inches (paper clips, books, pencils, sheets of paper, magazines, erasers, etc.)
- One ruler marked in inches for each student
- A narrow strip of paper, one foot long, for each student
- One yardstick for each student or group of students
- Scissors

Activity

Engage

Give each student a ruler and a narrow strip of paper that is approximately one foot long. Have the students locate the four-inch mark on their rulers. Point out that the distance from the zero mark to the four-inch mark is four inches. (Be sure that they start to count at the "zero" end of the ruler.) Have the students, without using their rulers, cut from their papers a piece that they judge to be four inches long. Have them measure the cut piece to see how close its lengths is to four inches. From the paper that remains, have the students cut a piece that they think is two inches long, then a piece that they think is three inches long.

Divide the students into three groups of approximately equal size. Have each group place all its estimates of four-inch strips in one pile, of two-inch strips in a second pile, and of three-inch strips in a third pile. The students should select the strip in each of their group's piles that they think is closest to the actual specified length. Next, they should measure the cut strips to check their selections. Finally, the groups' closest estimates should be compared and then measured.

Use tools to measure

Develop common referents for measures to make comparisons and estimates

Explore

Have the students use their rulers (marked in inches) to measure the lengths of paper clips, erasers, books, papers, magazines, staplers, and other objects that are less than twelve inches long. Describe the Estimation Challenge, in which the students estimate the lengths of objects in six "events" (i.e., six tasks on the worksheet) and compete for bronze, silver, and gold medals.

Give each student a copy of the blackline master "Estimation Challenge," and have the students complete the worksheet individually. They should then exchange papers in their groups and measure the actual objects with their rulers, to check the estimates. The students should then decide which members of the group made the three closest estimates for each "event." Have the students compare these estimates to determine a class winner for each event. Award gold, silver, and bronze medals to all the students in the groups with the first, second, and third greatest number of correct estimates, respectively. Be prepared with enough gold and silver medals in case of a two- or three-way tie.

Extend

As a follow-up to the worksheet activity, the students could estimate the lengths of the objects from the original "events" in half-inches and then measure to check their estimates.

To extend this activity, you could design another Estimation Challenge event involving larger objects. In the same way that you introduced the students to the ruler, introduce them to the yardstick, and point out the foot marks on it. Then have the students estimate the lengths, to the nearest foot, of large items in the classroom, in the gymnasium, or on the playground. Just as they did for the original events, the students can measure the objects (this time, with a yardstick) and determine the winning groups for each event.

If your yardsticks show only inches, not feet, use a marker, tape, or stickers to demarcate the feet.

Discussion

Through many activities like Estimation Challenge conducted over a long time, students develop expertise in estimating lengths. Estimation is an important skill that students should learn to apply to all measurable attributes—weight, capacity, and time, as well as length. Estimation is also important in the number strand. Students estimate "ballpark" figures when actual numbers are not required, and they estimate the number of objects in a set. They also use estimation to verify or check the results of computations.

Students should learn that all estimates require verification by measuring and that in reporting estimates they should use appropriate vocabulary—for example, *about* how long, *about* how heavy, and *about* how much time.

Grandma

Grades K–1

Summary

A dramatized situation helps students understand why standard units of length are important.

Goals

- Recognize the need for a standard unit of length
- Communicate measures of lengths

Prior Knowledge

- Measuring length with nonstandard units

Materials

- A three-foot length of ribbon (or string) for each group of students
- Extra string or ribbon (optional)
- A pair of scissors for each group of students
- A pencil and paper for each group
- A different nonstandard measuring unit for each group (e.g., small paper clips, small cubes, a small piece of ribbon, pencils of normal size, erasers of normal size, small sticky notes)
- Objects similar to those used as measuring units by the students, but of unusual sizes (e.g., large paper clips, large cubes, a very long and a very short pencil, a huge eraser, large sticky notes)
- Two toy telephones (optional)
- *How Big Is a Foot?* (Myller 1962)

Understand how to measure using nonstandard and standard units

Select an appropriate unit and tool for the attribute being measured

Use tools to measure

Develop common referents for measures to make comparisons and estimates

Activity

Engage

Divide the class into groups of three or four students. Supply each group with a three-foot length of ribbon (or string) and a nonstandard unit for measuring length. Have each group select one of the students to be measured. Instruct the other students in each group to wrap the ribbon or string around the selected student's waist, cut off the ribbon to a length equal to the size of the student's waist, measure the ribbon using the nonstandard unit, and record the measurement. Explain that the measurement will be used to make a belt for the student. Discuss with the students what they did when the length was not a whole number of units. Observe whether they rounded down or up; used language like *more than four, between four and five,* or *close to five;* or even used fractions like 4 1/2.

Explore

Tell the students that you will play the role of "Grandma," who lives out of town. You've agreed to make a belt for a costume for your grandchild, but he or she has to tell you how long to make it. Pick up one toy telephone (if one is available) and give the other to one of the students whose waists were measured. Have a "grandmotherly" conversation (see the sample conversation in fig. 2.12) with the "grandchild," eventually asking for the length of the belt.

Fig. **2.12.**

A sample grandmotherly conversation

Teacher:	Hi, Jeff. It's Grandma.
Jeff:	Hi, Grandma. How are you?
Teacher:	I'm fine, thank you. Jeff, your mom tells me that you are going to be an actor.
Jeff:	I'm going to play the part of Jack in my class's play.
Teacher:	Mom says you need a special belt. I told her I'd make it for you.
Jeff:	Thanks, Gram.
Teacher:	I'll need to know your size.
Jeff:	I measured my waist, and it's twenty paper clips long.
Teacher:	OK, Jeff. I'll make the belt and mail it to you before the play.
Jeff:	Great! Thank you, Grandma. Good-bye.

After you get off the phone, prepare to make the belt in front of all the children. Be outrageous in your interpretation of the measurement. For example, if the student says that the belt is four pencils long, pull out an amusingly long pencil and cut a piece of ribbon four times the length of that pencil. Pretend to wrap the belt and mail it to the grandchild. When the child receives the parcel and tries it on, the belt will be much too big. Act out the scenario with another group's representative, but this time deliberately misinterpret the size of the unit and make the belt much too small. Repeat the scenario with each group's representative, sometimes making a belt of reasonable size, but more often making one of unreasonable length. If appropriate, allow the groups to construct belts by following the other groups' directions.

Discuss with the students what could be done to ensure that Grandma knows exactly how long the belt should be. If the students do not mention it, introduce the concept of a *standard unit*, which would not be misinterpreted. Discuss standardized items that could be used, such as the length of a piece of standard loose-leaf paper. Alternatively, suggest a standard ruler.

Extend

If the book *How Big Is a Foot?* (Myller 1962) is available, read it to the children. The book tells the story of an apprentice to the king's carpenter who makes a bed for the queen by explicitly following the king's directions. The bed comes out the wrong size because the apprentice's foot, rather than the king's foot, was used to make the measurements. Discuss with the students how the story relates to Grandma's situation.

If the book is not available, have the students write a funny skit in which a parent's misunderstanding of a child's description of capacity leads to a ridiculous result. An example is a father selecting a box big enough to hold what his daughter has described as "four jars of marbles." The child's referent could be a mayonnaise jar, and the father's referent could be a baby-food jar.

Discussion

It is important that students understand why they use both nonstandard and standard units. One reason for using a nonstandard unit is to have a personal referent in order to make measurement meaningful. Knowing that the width of an object measures five of his or her hand spans means more to a young student than knowing that it is eighty centimeters wide. Another reason is to learn how to use whatever tools are available; adults, for example, will pace off a room when a yardstick is not available. Using a nonstandard unit allows the student to focus on the attribute being measured rather than be distracted by the size of a standard unit. The use of nonstandard units also helps students see the need for standard units, which is the point in this activity. The advantage of using standard units is to communicate measures unambiguously. Standard units are not intrinsically more useful as measurement tools than nonstandard units are.

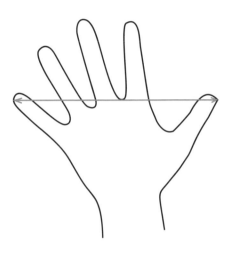

A hand span measures from the tip of the little finger to the tip of the thumb of a spread hand.

Calendar Logic

Grades 1–2

Summary

Students use deductive reasoning to determine the day of the week and the date on a calendar from written clues. The clues include the names of the days of the week; the digits that compose the date; terms for temporal relationships, such as *before*, *after*, and *between*; and the ordinal numbers first, second, third, and fourth.

Goals

- Use two or three clues to identify or eliminate seemingly possible solutions to a problem
- Use data presented in a calendar to solve problems

Prior Knowledge

- Identifying days and dates on a one-month calendar
- Adding sums to 30
- Subtracting from 30
- Counting by 2s and by 5s

Materials

- A copy of the blackline master "June" for each pair of students
- A copy of the blackline master "October" for each student
- A copy of a calendar for the current month for each pair of students

Activity

Engage

To help the students start thinking about relationships between the days of the week and the dates on a calendar, have them identify—

- the days of the week;
- the day just after Saturday;
- the second day of the week;
- the day just before Friday;
- the third week of the month;
- the days between Sunday and Thursday.
- the days of the weekend

Then pose the following questions:

- If today is Tuesday, what day of the week was yesterday?
- If yesterday was Saturday, what day of the week is today? Tomorrow?

"*When students use calendars or sequence events in stories, they are using measures of time in a real context.*"
(NCTM 2000, p. 104)

Point out to the students that *between* does not include the end points. Tell them, for instance, that the days between Sunday and Thursday are Monday, Tuesday, and Wednesday.

- If yesterday was Friday, what day of the week will the day after tomorrow be?

Ask the students to describe how they figured out the answers to the questions. Divide the class into pairs, and have the students in the pairs make up calendar questions to ask each other.

Explore

In this activity, the students use clues to identify a particular day of the week and date of the month. Some of the clues refer to the digits of a two-digit number. To set the stage for this activity, record several two-digit numbers on the board and talk about the digits of each number. For example, write the number 12 on the board and identify the digits 1 and 2 that form the number. Then write the number 15 on the board, and call on a student to identify the digits. (1 and 5) Follow the same procedure for the numbers 24, 19, 10, and 31.

Say, "I am thinking of one of these numbers." Point to 24, 19, 10, and 31. "When I add the digits of the number, the sum is 6. Which number am I thinking of?" (24) "I'm thinking of another number. When I add the digits, the sum is 10. What is the number?" (19)

Give each pair of students a copy of the blackline master "June." Call on students to—

- point to and name the days of the week;
- name the dates of all the Mondays in this June (7, 14, 21, 28);
- identify the number of Tuesdays in this June (5);
- identify the date of the day before June 20 (June 19);
- identify the day and date of the last day of June (Wednesday, June 30).

Have the students work in pairs to complete the June calendar problems on the worksheet. They can then complete the "October" worksheet individually.

Extend

Give each pair of students a copy of the calendar for the current month, and ask the students to create problems like the ones on the worksheets. Have the pairs present their problems to their classmates to solve. The student-created problems can be arranged in book form or laminated and placed in a "calendar logic" box in the mathematics center.

Discussion

As students solve calendar problems, they become familiar with temporal relationships and develop their spatial-temporal vocabulary. They also have opportunities to apply the counting and computation skills that they learn in the number strand.

Creating "calendar logic" problems enhances students' literacy skills while focusing their attention on aspects of the calendar and how the days of the week and their dates are related.

How Many in a ___?

Grades 1–2

Summary

Students estimate lengths and explore relationships between standard units of measure (both customary and metric) and common objects. They identify objects whose lengths are close to those of standard units and that can be used as referents when approximating or comparing lengths.

Goals

- Measure length using standard and nonstandard units
- Develop common referents for measures of length in order to make comparisons and estimates

Prior Knowledge

- Comparing lengths using nonstandard and standard units

Materials

- Rulers with both metric and customary units, yardsticks, and metersticks
- Common classroom materials, such as pencils, chalkboard erasers, and small and large paper clips
- Play money that is exactly the same length as U.S. dollar bills
- The following coins for each group of students: one quarter, ten dimes, ten nickels, and fifteen to twenty pennies. Play-money equivalents may be used if the coins have the same thickness as actual coins.

Activity

Engage

Tell the students that you are often out shopping or outside playing with children when you find that you need to measure something. Explain that you usually do not have a ruler with you, so you would like to be able to use a common item to help you approximate the lengths of different objects. Show the students a dollar bill, and tell them that since you usually have a dollar with you when you are shopping, you decided that you might be able to use that bill to help you measure different objects. Ask the students to estimate the number of dollar bills that would equal one foot and then check their estimates. Two dollar bills, placed end to end, are equal in length to a foot. (Each bill is six inches long.)

Explore

Explain to the students that since you were so successful in using a dollar bill to help you approximate measurements, you would like to explore how you might use coins, which you also usually have in your wallet, to help you make some approximate measurements. Separate the

Develop common referents for measures to make comparisons and estimates

students into groups of three or four, and give each group a ruler with both inches and centimeters marked, one quarter, ten dimes, ten nickels, and fifteen to twenty pennies. Tell the students that you would like them to determine if you could use the coins to approximate some common measures.

Let each group work independently on the task. The students might find that a quarter is about one inch in diameter or that a row of six nickels placed side by side is about five inches long. They might also discover that a stack of seven pennies or ten dimes is about one centimeter high. After the groups have had an opportunity to explore for a while, bring the entire class together to share their findings.

Extend

Collect an assortment of classroom supplies, such as large and small paper clips, unsharpened pencils, and sheets of computer paper, and place the items in a central location, where the children can see them and easily reach them. Tell the students that you would like to make a list of materials in the classroom that you and they could use for approximating measures when no standard tools are handy. The children might also use parts of their bodies, such as the length of a foot or the width of a finger. Let them know that you are especially interested in discovering equivalents for one inch, one centimeter, one foot, one meter, and one yard. Have them work in the small groups to find referents for the indicated measures. After the students have had some time to investigate, bring them together to share their information, and record it in a class chart like the one in figure 2.13.

Measure	Referent
About one centimeter	The width of a large paper clip, the height of a stack of 7 pennies or 10 dimes
About one inch	The width of a quarter
About one foot	Two dollar bills placed end to end; the length of a sheet of paper
About one meter	The height of a doorknob from the floor. The width of a doorway, including the door jambs
About one yard	(Possibly) The height of the tallest student's waist from the floor

Fig. **2.13.**

A classroom chart of the measures of common objects

Help the students understand that some of these referents will change over time. For example, in the sample chart, the height of the tallest student's waist from the floor is suggested as a referent for one yard. Ten years from now, that height may be much greater than one yard. Barring an overhaul of our paper currency, however, two dollar bills will remain a good referent for one foot.

Discussion

This activity builds on students' understanding of common units of measure and gives them referents to use in approximating lengths. Take the opportunity to discuss with the students the accuracy of the measurements given by different tools. Explain why accurate measuring tools are needed for some tasks, whereas good approximations suffice for other tasks.

For example, if Maria used her hand span to measure the ribbon for a fancy bow to decorate a present, she would get a measure of ribbon that would be sufficient for her purpose, since the bow could be smaller than she had intended it to be and still make an attractive decoration. If she wanted to buy ribbon to decorate the hem of her dress, however, she might want to use a precise tool, such as a measuring tape, to ensure that she bought enough ribbon to go around the entire hem.

Cover Up

Grades 1–2

Summary

Students compare areas as they figure out which cutout shape is larger. They learn to compare by superimposing one figure onto another; when that approach doesn't work, they cut one figure into pieces that can fit on top of the other figure. Comparing the areas of objects that cannot be cut and reassembled, such as a rectangular cookie pan and a circular pizza pan, helps students recognize the need for using standard units.

Goals

- Compare areas by superimposing one figure onto another
- Compare areas by cutting one shape and using the pieces to cover another shape
- Compare areas by counting and comparing the numbers of units needed to cover two figures

Prior Knowledge

- Recognizing shapes

Materials

- Large shapes cut from 8 1/2-inch-by-11-inch colored construction paper (circles, squares, rectangles, triangles, hexagons, pentagons, kites)—at least one for each student and about fifteen others for use in a demonstration
- One pair of scissors for each pair of students
- Approximately seventy-five construction-paper or tile squares, one inch on each side, for each pair of students
- One hundred construction-paper or tile circles, one inch in diameter
- A metal cookie sheet and a circular metal (pizza) pan

Activity

Engage

Show the students two shapes, one of which covers the other completely. Ask the students to identify the one that is larger (smaller) and to tell the class how they know it is larger (smaller). Continue with several more comparisons. Be sure that the students learn that one way to compare areas is by superimposing one shape onto the other.

Explore

Show the students two polygons (closed shapes with sides that are line segments) that cannot be compared by superimposition—that is, neither shape completely covers the other (see fig. 2.14). Invite suggestions about how the shapes might be compared to determine which is

Apply appropriate techniques, tools, and formulas to determine measurements

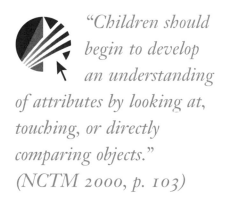

"Children should begin to develop an understanding of attributes by looking at, touching, or directly comparing objects." (NCTM 2000, p. 103)

Fig. **2.14.**

A triangle that does not completely
cover the square

larger (smaller). If the students do not suggest cutting one shape apart and fitting the pieces onto the other shape, you should do so. Call on students to identify the shape to be cut, how it should be cut, and how the cut pieces should be arranged on the other shape. Follow the students' suggestions, and have them identify the shape that is larger or tell if both shapes are about the same size.

Follow the same procedure for other pairs of shapes; use polygons for the first five comparisons, and include circles along with polygons in later comparisons.

Extend

Show the students the rectangular cookie sheet and the circular pizza pan. Ask, "Which pan is larger?" Since superimposing doesn't work and the pans cannot be cut apart, encourage the students to suggest ways to compare the two shapes. The students may suggest tracing the shapes of the two pans onto paper and then superimposing one shape onto the other or cutting a shape and fitting its pieces onto the other shape. Follow through on their suggestions.

Next, show the students the small (one-inch) squares and circles. Demonstrate how to use the squares to cover the rectangular pan, count the small squares, and use that number of squares as a measure of the area of the pan. Do the same for the pizza pan. (Some squares will have to be cut to fit the pans.) Record the areas of the two pans, and ask the students which pan is larger (i.e., measures more square units) and which pan is smaller (i.e., measures fewer square units).

Follow the same procedure in covering the pans with the small circles. Point out that the circles should not overlap. Talk with the students about which shape—the small square or the small circle—is better suited to measuring the pans.

Divide the students into pairs. Give each pair two of the large shapes that were not cut apart in the "Explore" part of the activity and approximately seventy-five small squares. Have the students use the small squares to measure and compare the areas of the shapes.

Discussion

The activities in Cover Up lead students to recognize the need for a standard unit for measuring area. Before teachers introduce the unit square, however, their students should have many opportunities to use different polygons to cover a surface, and they should be able to describe why certain polygons fit together better than others do. You might want to use some tessellation activities with pattern blocks in conjunction with Cover Up to reinforce the idea of shapes "fitting together."

Which Unit Did I Use?

Grade 2

Summary

Given different measuring tools and a list of items, students work in groups to measure the lengths of the items. Although the items on the list are the same for each group, the designated unit may be a centimeter (cm), a decimeter (dm), or a meter (m). When the students have completed all the measurements, they announce the number of units that the groups found for each item without indicating the units used. The other students then try to identify the unit that each group used.

Goals

- Recognize the inverse relationship between the size of a measurement unit and the numeric measure
- Recognize the most appropriate units and tools to measure different lengths

Prior Knowledge

- Using customary and metric units to measure length

Materials

- Metersticks
- Centimeter rulers
- Decimeter rods (In most sets of base-ten blocks, the tens are ten centimeters long; measure the rods in the sets you are using to be certain.)
- Three pieces of string or heavy yarn of the same weight; they should be different colors, if possible, and have the following lengths: five centimeters, five decimeters, five meters
- Pencils
- For each group of students, a list of objects in the classroom that the students could measure in centimeters, decimeters, or meters. Identify the part or dimension of each item to be measured, perhaps with a drawing. Include a variety of lengths. (See the list in fig. 2.15.)

Select an appropriate unit and tool for the attribute being measured

Use tools to measure

Activity

Engage

Display the three pieces of string or yarn rolled into separate balls and a meterstick, a centimeter ruler, and a decimeter rod. Call on students to identify the length of each measuring tool. Have the students identify the string they think is the longest (or shortest) and explain why they think so. Then have them identify the strings by length. Which do they think is five centimeters? Five decimeters? Five meters? Extend the three strings. Call on a student to demonstrate how to measure the length of the string thought to be five centimeters long. Before

Fig. **2.15.**

A list of objects that students can measure

Write the number and the unit for each measurement.

"*If some students measure the width of a door using pencils and others use large paper clips, the number of paper clips will be different from the number of pencils.*" (NCTM 2000, p. 105)

The applet Which One? on the CD-ROM is designed to continue students' investigation of the relationship between the size of a unit and the numeric measure.

he or she measures, have the class decide which unit and measuring tool should be used. Call on different students to find the lengths of the other two strings, and have the class decide on the units to be used to measure them.

Explore

Divide the students into small groups. Give each group one of each kind of measuring tool and the list of items to measure. Encourage the groups to discuss which unit they will use (cm, dm, or m) to measure each item, and have them record the unit on the list. Note that it is unlikely that any length will measure an exact number of units, so the students will need to decide what to do when the measurement cannot be expressed in a whole number of units. For example, they may count to the nearest unit or use phrases such as *a little bit more than five centimeters*.

When all the groups have completed the task, have them share the number of units they found for the first item on the list but without indicating the specific units used. The other students should then try to identify the unit used by the other groups. Repeat the process for each item on the list.

Extend

Discuss how the students might choose units and tools that match the size of the object they wish to measure. Ask questions such as

"Would you use centimeters or meters to measure the width of your hand?" "Why?" Guide the students to see the inverse relationship between the size of the unit and the numeric measure: the larger the unit, the fewer of them, and the smaller the unit, the more of them. For example, a bookcase might be two meters long (big unit, small numeric measure). If it were measured in centimeters, the same bookcase would be 200 centimeters long (small unit, large numeric measure).

Discussion

Explain that the lengths of objects can be measured with large or small linear units but that the number of units will be different for units of different sizes. It is generally more efficient to measure small objects with small units and large objects with large units. The size of the object to be measured and the sizes of the available units help determine which tool would be the most appropriate. Although a crayon can be measured with a meterstick, it would be easier to measure it with a shorter centimeter ruler. Similarly, it would be easier to use a meterstick and measure the height of a chalk tray to the nearest meter than to use a decimeter rod and measure the same height to the nearest decimeter. It should be noted, however, that carefully measuring to the nearest whole unit in smaller units yields a more precise measurement than measuring to the nearest whole unit in larger units.

Fit the Facts

Grade 2

Summary

The problems in Fit the Facts are stories with blanks that represent missing measures. A list of "facts" (numerical measures) is given for each story. The students must use their measurement sense to use the given numbers to fill in the blanks so that the story makes sense.

Goals

- Use known referents to estimate measures
- Develop measurement sense

Prior Knowledge

- Recognizing such standard units as miles, pounds, inches, and quarts

Materials

- A transparency copy of the blackline master "All about Jamal"
- An overhead projector
- A copy of the blackline master "All about Marie" for each student or pair of students

Activity

Engage

To begin a class discussion, ask, "What is a fact?" Allow time for the students to offer various ideas. Often students will suggest that it is something that you know or can find out. Encourage the students to share facts about themselves. For example, they may suggest where they live, the number of people in their families, or their likes and dislikes. To further prepare the students for the activity, ask, "Do you know a measurement fact about yourself?"

Explore

Say to the students, "I will tell you a story, but some of the numbers are missing." On the overhead projector, display "All about Jamal." Read the problem to the class, saying, "blank" when appropriate. Have the students join you in a second reading of the problem.

Divide the students into pairs, point to the numbers in the "Numbers" box, and allow some time for the students to talk in pairs about where the numbers might fit into the story. Then ask, "Which number did you decide on first?" Note that the numbers do not need to be placed in the blanks in the order of the story. In fact, you should encourage the students to proceed in any order they wish. Discuss the students' ideas, and accept their reasonable suggestions until all the blanks have been filled in. Then call on a student to read the completed story aloud (see the margin).

"Estimation activities are an early application of number sense; they focus students' attention on the attributes being measured, the process of measuring, the sizes of units, and the value of referents."
(NCTM 2000, p. 106)

All about Jamal

Jamal is in grade ___2___ .

He is __49__ inches tall.

His pencil is _6_ inches long.

Jamal's dog, Sammy, weighs ___27___ pounds.

The students are likely to refer to their own heights in order to decide what Jamal's height might be. If they don't, ask, "Do you know how tall you are? Are you taller or shorter than Jamal? How does knowing how tall you are help you decide which number to choose for Jamal's height?" Similarly, the lengths of the students' pencils can help them identify the correct length of Jamal's pencil.

Extend

Distribute a copy of "All about Marie" to each student or pair of students to be completed individually or together. When the students have completed their work, have them discuss their decisions and give reasons for their placement of the number "facts."

Discussion

In Fit the Facts, students reason mathematically and communicate their thinking while developing their measurement sense. When students refer to measurements that they know, such as the weight of a baby sister or the length of a baseball bat, they are learning to use referents to estimate measures. This important skill is used frequently in daily life.

 *"Discourse builds students' conceptual and procedural knowledge of measurement and gives teachers valuable information for reporting progress and planning next steps."*
(NCTM 2000, p. 103)

NAVIGATING *through* MEASUREMENT

Looking Back and Looking Ahead

In prekindergarten through grade 2, students compare and order objects or events on the basis of an attribute: length, weight, capacity, area, or time. They compare lengths directly and indirectly. They learn to differentiate heavy and light by feel and to use a pan balance to compare objects of similar weight. They compare the capacities of two containers by pouring the contents of one into the other. Such activities help children attend to specific attributes and to develop comparative language.

Once students have developed a well-grounded understanding of attributes through comparison activities, they are ready to explore the use of units and tools for measuring. They use a variety of nonstandard and standard units to measure length, weight, capacity, area, and time. In this grade band, the emphasis is on linear measurement. In grades 3–5, the study of measurement is extended to a deeper examination of area and volume, and it includes the measurement of parts of shapes or objects (e.g., angles, perimeters, circumferences).

Although the emphasis is on nonstandard units in prekindergarten through grade 2, standard units are also introduced. Through a variety of explorations, students begin to understand the general process of measurement: identifying an attribute and a unit and then comparing the unit to the object or event in order to assign a number. A more formal exploration of customary and metric units—including simple conversions within a particular system of measurement—takes place in grades 3 through 5.

In prekindergarten through grade 2, students begin to measure by using multiple copies of the same unit. Eventually they are able to use one unit repeatedly. Over time, students improve their abilities to place units end to end without overlap, and their accuracy in measuring tasks is expected to improve in grades 3–5. Although tools such as the ruler, pan balance, and calendar are introduced in prekindergarten through grade 2, counting units and estimating are the primary measurement activities for the youngest students. In grades 3–5, students begin to use formulas.

Estimation is an important component of the measurement process; it helps students become more familiar with units and with the process of measurement. Estimation, along with activities involving the use of units of different sizes to measure the same object, offers students an informal way to explore the relationship between the size of units and the number of units. In grades 3–5, students formalize this inverse relationship.

The prekindergarten–grade 2 measurement curriculum lays the foundation for the further development of many fundamental ideas of measurement in grades 3–12; it also affords students opportunities to apply what they are learning in the number, algebra, geometry, and data and probability strands. Experiences with measurement concepts will lead to success in the learning of more-complex ideas, not only in measurement but also in number, algebra, geometry, statistics, and probability.

NAVIGATING *through* MEASUREMENT

PRE-K–GRADE 2

Appendix

Blackline Masters and Solutions

Body Balance

Name _____

Draw a line to the correct hand for each item.

1.　　　　2.　　　　3.　　　　4.

5. Draw something in each box to make the body balance correct.

Possible or Not?

Name _____

Circle the pictures that are *not* possible.

1.

2.

3.

4.

5. Draw objects in the boxes so that the picture is *not* possible.

Scavenger Hunt List

Name _____

1.   is lighter than a

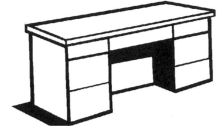

2.   is longer than a

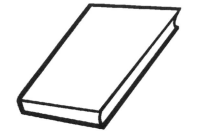

3.   holds less than a

4.   is heavier than a

5.   is shorter than a

6.   holds more than a

Estimation Challenge

Name _____

Use your estimation skills in these six Estimation Challenge "events."

1. Point *A* is one end point of a 4-inch line segment. Draw and label point *B* to show where the line segment should end.

   A •

2. About how many inches long is the line segment from point *A* to point *B*?

   A •————————————• *B*

3. About how many inches long is the line segment from point *C* to point *D*?

4. Draw a line segment that you estimate to be 5 inches long.

5. About how many inches long is a brand-new piece of chalk? _____

6. About how many inches long is your mathematics book? _____

Estimation Challenge Medals

Copy the "medals" onto card stock or heavy paper. Use crayons or other markers to color the medals appropriately. Punch out holes where indicated. Thread ribbons through the holes, and tie the ends of the ribbons. The number of medals you will need depends on the sizes of the three groups of students. Be prepared with enough gold or silver medals in case of a two- or three-way tie.

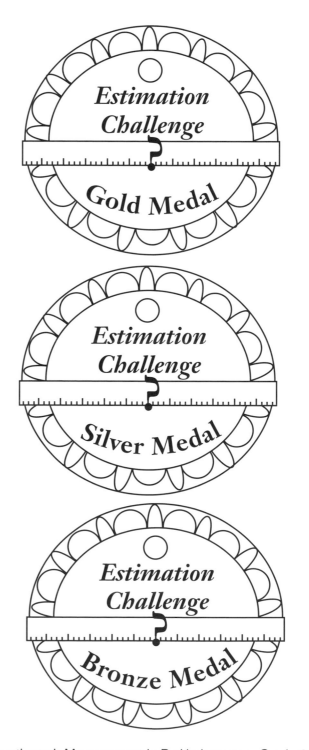

June

Names _____

_____

Use the clues and the calendar for June to name the day of the week and the date of the month.

1. Clues
   - One of the digits of the date is a 1.
   - The other digit of the date is less than 1.

   Day of the week _____   June _____

2. Clues
   - The sum of the two digits of the date is 8.
   - The date is a number you say when you count by 2s.

   Day of the week _____   June _____

3. Clues
   - The day is not on a weekend.
   - The date is a number you say when you count by 5s.
   - The sum of the digits of the date is 6.

   Day of the week _____   June _____

4. Clues
   - The date is after June 20.
   - The ones digit of the date is twice the tens digit.

   Day of the week _____   June _____

JUNE

Sunday	Monday	Tuesday	Wednesday	Thursday	Friday	Saturday
		1	2	3	4	5
6	7	8	9	10	11	12
13	14	15	16	17	18	19
20	21	22	23	24	25	26
27	28	29	30			

Navigating through Measurement in Prekindergarten–Grade 2

October

Name _____

Use the clues and the October calendar to name the day of the week and the date of the month.

1. Clues

    • The date is after October 26.
    • The name of the day begins with *S*.

    Day of the week _____   October _____

2. Clues

    • The day is the last school day in the week.
    • One of the digits of the date is 3.
    • The day is not in the last week of the month.

    Day of the week _____   October _____

3. Clues

    • You say the date when you count by 2s.
    • One of the digits of the date is double the other.
    • The day is the first school day of the week.

    Day of the week _____   October _____

4. Clues

    • The day is between two days of the week whose names begin with the same letter.
    • The sum of the digits of the date is 5.

    Day of the week _____   October _____

OCTOBER

Sunday	Monday	Tuesday	Wednesday	Thursday	Friday	Saturday
				1	2	3
4	5	6	7	8	9	10
11	12	13	14	15	16	17
18	19	20	21	22	23	24
25	26	27	28	29	30	31

Navigating through Measurement in Prekindergarten–Grade 2

All about Jamal

Jamal is in grade _____.

He is _____ inches tall.

His pencil is _____ inches long.

Jamal's dog, Sammy, weighs _____ pounds.

Numbers
49   **2**
6   **27**

All about Marie

Name _____

Marie is _____ year(s) old.

Each day after school, she rides her

bicycle for _____ mile(s).

She drinks _____ pint(s)

of water after her ride.

On Saturdays, Marie plays baseball.

Her bat is _____ inch(es) long.

Numbers	
26	**8**
3	**1**

Marie has a new baby brother,

Raphael.

He was born _____ days, or

_____ weeks, ago.

Raphael is _____ inches long

and weighs _____ pounds.

Numbers	
19	**2**
7	**14**

Navigating through Measurement in Prekindergarten–Grade 2

Solutions for the Blackline Masters

Solutions for "Body Balance"

1–4. See the illustration.

5. Many answers are possible. A student could have drawn two books of similar size, one on each side.

Solutions for "Possible or Not?"

1–4. See the illustration.
5. Many answers are possible. One example is a balloon in the figure's right hand and a full backpack in its left hand.

Solutions for "Scavenger Hunt List"

1–6. Answers will vary.

Solutions for "Estimation Challenges"

1. Have the students measure to determine how close the line segments are to four inches.
2. The actual line segment is three inches long.
3. The actual line segment is four inches long.
4. Have the students measure to determine how close the line segments are to five inches.
5. Answers will vary. Measure the length of the chalk you use in your classroom.
6. Answers will vary. Measure your students' mathematics textbook.

Solutions for "June"

1. Thursday, June 10
2. Saturday, June 26
3. Tuesday, June 15
4. Thursday, June 24

Solutions for "October"

1. Saturday, October 31
2. Friday, October 23
3. Monday, October 12
4. Wednesday, October 14

Solutions for "All about Marie"

Marie is ___8___ year(s) old.

Each day after school, she rides her bicycle for ___3___ mile(s).

She drinks ___1___ pint(s) of water after her ride.

On Saturdays, Marie plays baseball.

Her bat is ___26___ inch(es) long.

Marie has a new baby brother, Raphael.

He was born ___14___ days, or ___2___ weeks, ago.

Raphael is ___19___ inches long and weighs ___7___ pounds.

References

Greenes, Carole. "Ready to Learn: Developing Young Children's Mathematical Powers." In *Mathematics in the Early Years*, edited by Juanita V. Copley, pp. 39–47. Reston, Va.: National Council of Teachers of Mathematics; Washington, D.C.: National Association for the Education of Young Children, 1999.

Hiebert, James. "Why Do Some Children Have Trouble Learning Measurement Concepts?" *Arithmetic Teacher* 31 (March 1984): 19–24.

Kellogg, Steven. *Much Bigger than Martin*. New York: Dial Press, 1976.

Lang, Frances Kuwahara. "What Is a 'Good Guess,' Anyway? Estimation in Early Childhood." *Teaching Children Mathematics* 7 (April 2001): 462–66.

Myller, Rolf. *How Big Is a Foot?* Bloomfield, Conn.: Atheneum, 1962.

National Council of Teachers of Mathematics (NCTM). *Principles and Standards for School Mathematics*. Reston, Va.: NCTM, 2000.

Outhred, Lynne N., and Michael C. Mitchelmore. "Young Children's Intuitive Understanding of Rectangular Area Measurement." *Journal for Research in Mathematics Education* 31 (March 2000): 144–67.

Piaget, Jean, Bärbel Inhelder, and Alina Szeminska. *The Child's Conception of Geometry*. New York: Basic Books, 1960.

Ryder, Joanne. *The Snail's Spell*. New York: Puffin Books, 1988.

Tompert, Ann. *Just a Little Bit*. Boston: Houghton Mifflin Co., 1993.

Wertsch, James V. *Mind as Action*. New York: Oxford University Press, 1998.

Suggested Reading

Friederwitzer, Fredda J., and Barbara Berman. "The Language of Time." *Teaching Children Mathematics* 6 (December 1999): 254–59.

Greeenes, Carole, Linda Schulman Dacey, and Rika Spungin. *Hot Math Topics: Spatial Sense and Measurement (Grade 2)*. Parsippany, N.J.: Dale Seymour Publications, 2001.

Hiebert, James. "Cognitive Development and Learning Linear Measurement." *Journal for Research in Mathematics Education* 12 (May 1981): 197–211.

Hildreth, David J. "The Use of Strategies in Estimating Measurements." *Arithmetic Teacher* 30 (January 1983): 50–54.

Kamii, Constance. "Measurement of Length: The Need for a Better Approach to Teaching." *School Science and Mathematics* 97 (March 1997): 116–21.

Liedtke, Werner W. "Measurement." In *Mathematics for the Young Child*, edited by Joseph N. Payne, pp. 228–49. Reston, Va.: National Council of Teachers of Mathematics, 1990.

Lindquist, Mary Montgomery. "The Measurement Standards." *Arithmetic Teacher* 37 (October 1989): 22–26.

Lubinski, Cheryl A., and Diane Thiessen. "Exploring Measurement through Literature." *Teaching Children Mathematics* 2 (January 1996): 260–63.

McClain, Kay, Paul Cobb, Koeno Gravemeijer, and Beth Estes. "Developing Mathematical Reasoning within the Context of Measurement." In *Developing Mathematical Reasoning in Grades K–12*, 1999 Yearbook of the National Council of Teachers of Mathematics, edited by Lee V. Stiff, pp. 93–106. Reston, Va.: National Council of Teachers of Mathematics, 1999.

Pike, Christopher, and Michael Forrester. "The Influence of Number-Sense on Children's Ability to Estimate Measures." *Educational Psychology* 17 (December 1997): 483–501.

Porter, Jeanna. "Balancing Acts." *Teaching Children Mathematics* 1 (March 1995): 430–31.

Rhone, Lynn. "Measurement in a Primary-Grade Integrated Curriculum." In *Connecting Mathematics across the Curriculum*, 1995 Yearbook of the National Council of Teachers of Mathematics, edited by Peggy A. House, pp. 124–33. Reston, Va.: National Council of Teachers of Mathematics, 1995.

Richardson, Kathy. "Too Easy for Kindergarten and Just Right for First Grade." *Teaching Children Mathematics* 3 (April 1997): 432–37.

Ruggles, JoLean, and Barbara Sweeney Slenger. "The 'Measure Me' Doll." *Teaching Children Mathematics* 5 (September 1998): 40–44.

Schwartz, Sidney L. "Developing Power in Linear Measurement." *Teaching Children Mathematics* 1 (March 1995): 412–16.

Stephan, Michelle, Paul Cobb, Koeno Gravemeijer, and Beth Estes. "The Role of Tools in Supporting Students' Development of Measuring Conceptions." In *The Role of Representation in School Mathematics*, 2001 Yearbook of the National Council of Teachers of Mathematics, edited by Albert A. Cuoco, pp. 63–76. Reston, Va.: National Council of Teachers of Mathematics, 2001.

Wilson, Patricia S., and Ruth E. Rowland. "Teaching Measurement." In *Research Ideas for the Classroom: Early Childhood Mathematics*, edited by Robert J. Jensen, pp. 171–94. New York: Macmillan Publishing Co., 1993.

Children's Literature

Allen, Pamela. *Who Sank the Boat?* New York: Coward, McCann, and Geoghegan, 1983.

Anno, Mitsumasa. *The King's Flower.* New York: Collins, 1979.

Anno, Mitsumasa, and Raymond Briggs. *All in a Day.* New York: Philomel Books, 1986.

Carle, Eric. *The Very Hungry Caterpillar.* New York: Putnam, 1969.

Hoban, Tana. *Big Ones, Little Ones.* New York: Greenwillow Books, 1976.

———. *Is It Larger? Is It Smaller?* New York: Greenwillow Books, 1985.

Leedy, Loreen. *Measuring Penny.* New York: Henry Holt, 1997.

McMillan, Bruce. *Super, Super, Superworlds.* New York: Lothrop, Lee & Shepard, 1989.

Silverstein, Shel. "One Inch Tall." In *Where the Sidewalk Ends: The Poems and Drawings of Shel Silverstein*, p. 55. New York: Harper and Row, 1974.